RECORDING UNHINGED

RECORDING UNHINGED

CREATIVE AND UNCONVENTIONAL MUSIC RECORDING TECHNIQUES

Sylvia Massy
with Chris Johnson

Online content can be downloaded here:

https://resources.rowman.com/download?isbn=9781495011276&media=recordingonlinemedia

Hal Leonard Books
An Imprint of Hal Leonard Corporation

Copyright © 2016 by Sylvia Massy

All rights reserved. No part of this book may be reproduced in any form, without written permission, except by a newspaper or magazine reviewer who wishes to quote brief passages in connection with a review.

Published in 2016 by Hal Leonard Books
An Imprint of Hal Leonard Corporation
7777 West Bluemound Road
Milwaukee, WI 53213

Trade Book Division Editorial Offices
33 Plymouth St., Montclair, NJ 07042

Illustration credits can be found on pages 243–244, which constitute an extension of this copyright page.

Printed in the United States of America

Book design by Damien Castaneda

Library of Congress Cataloging-in-Publication Data
Names: Massy, Sylvia. | Johnson, Chris.
Title: Recording unhinged : creative and unconventional music recording techniques / Sylvia Massy with Chris Johnson.
Description: Montclair, NJ : Hal Leonard Books, 2016.
Identifiers: LCCN 2015044344 | ISBN 9781495011276
Subjects: LCSH: Popular music--Production and direction. | Sound-
-Recording
  and reproducing.
Classification: LCC ML3790 .M353 2016 | DDC 781.49--dc23
LC record available at http://lccn.loc.gov/2015044344

ISBN 9781495011276

www.halleonardbooks.com

CONTENTS

1	**CHAPTER 1.** **ADVENTURES IN RECORDINGLAND**
1	Disclaimer: Use This Book at Your Own Risk!
2	The Operator's Point of View
2	A Banquet for the Senses
4	Unbuckle Up
5	The Importance of NOT Having a Plan
5	Happy Accidents
7	Ross Robinson on Encouraging Mistakes
9	**CHAPTER 2.** **ENGINEERS AND DEMOLITION EXPERTS**
9	Fearless Recording!
9	Elliot Scheiner on Stupid Ideas That Work
10	Let It Bleed!
14	Commitment. Be Brave
14	Rainbows, Unicorns, and the Agony of Defeat
15	Technically, It's All Noise
15	The Right Gear Makes It Easy
17	The Era of Giants
18	Neve
19	Trident
20	API

22	SSL
22	OK, Now What?
22	Creaming the Mic Pres
23	Untamed Double Compression
24	Creepy Old Gear
24	Sta-Level
26	Collins
26	RCA
27	The Mish-Mashers
29	My Bestest Friend Ever
30	The Studio Standards
31	Do You Even Need Compression?
31	Spaces, Places, and Trains
32	Clearwell Castle
32	Castle Röhrsdorf
32	The Majestic Taj Mahal
32	Al Schmitt's Taj Mahal Experience
33	Teatro
33	RadioStar Studios
34	AIR Studios
35	Headley Grange
37	Real World Studios
38	Wayne Coyne's Parking Lot Experiment
39	"Station to Station"
39	Justin Stanley's On-the-Rails Odyssey
42	Bob Ezrin on David Gilmour's Astoria
42	Bob Clearmountain on Recording on Top of a Mountain

CHAPTER 3.
ALL THE WORLD'S A STUDIO

45	Opening a Scene
45	The Bravery of Live-to-Two-Track
47	Ultimate EQ Commitment
48	Analog When You Don't Have To
49	Other Analog Alternates
50	Record with Whatever!!!
51	Loud Noises and Smashing Things
54	Studio Pals

CHAPTER 4.
VOCALS

57	Svengali Vocalist Mind Control
59	Warm Up or You Might Fall Down
59	I'm Picking Up Good Distractions
62	Lights, Camera, Vocals!
62	Naked and Vocal About It
64	Go Over the Top
64	A Punch in the Gut
66	Performance over Perfection
68	The Trouble with Harmony
69	And Now He's Sick, But Of Course
70	The Posse Is in the House
70	Remedies Without Pharmacies
71	Pop-Up Vocal Booth
71	Sylvia's Secret Vocal Compression
72	Headphone Mix to Control Vocal Pitch
74	Headphone Levels to Control Performance
75	Modern Microphone Champions
76	The Oddballs
77	Matt Wallace on Singing Into a Wall
78	The Color and the Shape
78	To Reverb or Not to Reverb?
78	Comfort/Discomfort
79	Pissing Off the Singer
80	Handheld Versus Suspended Microphones
82	Jumping in the Deep End
82	Singing in an Aquarium
82	Caution About Power Hazards
83	Singing Through a Snare Drum
83	Singing Through a Fan
84	Singing Upside Down
85	How to Build a Telephone Mic
86	Have a Drink on Me

CHAPTER 5.
BASS

89	Fingers, Picks, and What?
89	Sexy, Sexy Bass
90	The Magic Combinations

91	Low-down Microfunken and EQ
92	Absolute Distortion on Everything
93	Using Guitar Amps for Bass
95	Stringing Them Along
95	String Cheese and Ham
97	More Than Electric Basses

CHAPTER 6.
DRUMS
99

99	What Is the Big Deal?
100	Think Like a Drummer
101	Tricks for Happy Drum Sounds
102	Use Your Heads
102	Learn to Tune Drumheads
103	Tuning Toms to the Key of the Song
103	Want Exciting Drum Sounds? Change Heads Often
104	Removing Bottom Heads
105	Dirty, Dirty Cymbals
105	Sylvia's Salted Cymbal Recipe
107	Don't Point That Stick at Me
107	What Type of Beater?
108	Phase Is God
109	Building a Sub Mic for Your Kick
110	Jack Joseph Puig on Sub Mic Dangers
111	One Mic, Two Personalities
111	Sexy Positions
111	Miking the Kick Beater
112	Damping Drums with Scissors and Bicycle Tubes
113	The Sure Things
115	Miking the Snare Port
115	Going Shotgun
116	Nothing Is Perfect
118	Weird Drums
120	Prepared Drums and Alternates

CHAPTER 7.
GUITAR
123

123	Electric Attitude
124	Curious Wood and Wire
125	Amplifiers Unhinged

125	The New Weirdos
127	Tuning a Ham-Fisted Player
128	Tiny Amps, Big Sound
130	A Pantry Full of Microphones
131	Phase Flip and Commit!
132	Microphone Robots
132	The Split-Amp Technique
134	Indecent Guitar EQ
134	Moving Targets
135	Evil Little Boxes
136	Be the Guitar Player's Foot
137	More Than Guitar
138	Tasty Talkbox
138	Control Room Feedback Generator
140	Matt Wallace on Building Weird Shit
141	Justin Stanley on Getting Sweet Distortion on a Direct Guitar

CHAPTER 8.
PIANO AND ORGAN
143

143	Upright Piano as a 19th-Century Entertainment Center
144	The Grander, the Better?
145	Peculiar Miking
145	Lid Reflections
145	Grandiose Upright
146	Prepared Piano and Other Extended Piano Techniques
146	Playing Piano with Something Other Than Hammers and Keys
146	Tack Piano
147	Rick Rubin's Single-Note Impact Piano
147	Clav, Rhodes, Harpsichord, and Wurli
148	Hammond Cheese
150	Show Me Your Pipes

CHAPTER 9.
STRINGS, HORNS, AND ORCHESTRA
153

153	Keeping It Old-School
155	Decca Tree

157	Ultimate Mics
160	And What Else?

165 CHAPTER 10. KEYS, SYNTHS, AND SAMPLERS

165	Early Mechanical Beasts
170	Patching Through Analog Synths
172	Tape- and Disc-Based Samplers
173	The Weirder, the Better
174	Julian Colbeck on Memory "Fossilization"

177 CHAPTER 11. PERCUSSION AND OTHER NOISE

177	Slaps, Dings, and Rattles
178	Michael Beinhorn on the Soundgarden Spoons
181	The Art of Pandemonium
184	Downright Wack-a-doo
185	Playing a Motorcycle Solo

187 CHAPTER 12. PRODUCTION APPROACH

187	What Does a Producer Do?
189	The Pioneers
189	Raymond Scott
190	Spike Jones
190	John Cage
190	Phil Spector
190	Sylvia Vanderpool-Robinson
191	Karlheinz Stockhausen
193	The Rise of the Independents
194	Lee "Scratch" Perry
195	The Musician Producers
196	Linda Perry on Channeling Music
197	The Rebel Producers
197	Chris Thomas
197	Konrad "Conny" Plank
197	Brian Eno
198	Mike Patton

199	The Mad Scientists
199	Rick Rubin
200	Song Structure and Content
200	Preproduction
201	Intros and Outros
202	Pressure-Cooker Homework for the Band
202	Write for Another Artist to Get Out of a Rut
203	Click Decisions
206	Hooks, Hooks, Hooks
206	Lyrical Fodder
207	Blending the Palette
209	The Importance of "Negative Space"
209	Get the Damn Record Finished, Set Limits
210	The Good, the Bad, the Ugly
211	Breakthrough Production Ideas
213	Jumping the Shark

CHAPTER 13. MIXING

Starting on page 215

215	Perfection Mixing
216	Repair Mixing
217	Mixing by Committee
217	Adventure Mixing
221	Big Dumb Tom Panning
221	Left, Right, Center
222	Re-Amping with the "Double-DI" Technique
223	Matt Wallace on Re-Amp Chamber Reverbs
225	Ross Hogarth on the Passive Transformer "Iron Giant"
226	Paul Wolff's "Hey-Fix" Mix Clarifier
228	Building Your Own Cooper Time Cube
229	Tchad Blake on Tom Waits's Splattered Mix
231	When Is a Mix Finished?

235	**ACKNOWLEDGMENTS**
237	**INDEX**
243	**ILLUSTRATION CREDITS**

Figure 1-1.

1

ADVENTURES IN RECORDINGLAND

Figure 1-2. Kaiser's Orchestra on Preikestolen cliff in Norway.

DISCLAIMER: USE THIS BOOK AT YOUR OWN RISK!

Here's a question: Why walk when your car is sitting in the driveway? Walking certainly isn't the fastest mode of travel. But when we use our feet, we often take a different route to our destination. It usually takes longer, but we might enjoy the scenery and weather, and we could even see something along the way we didn't expect. We reconnect with our surroundings and interact with people.

Modern recording can be like driving a Buick. You're comfortably controlling things from behind the glass—but perhaps that comfort lulls one into mediocrity. Fear generates adrenaline, and

Hans Zimmer on Taking Risks

"Film director Christopher Nolan gave me a watch at the end of the *Interstellar* film project, and on the back it says, 'This is no time for caution.' And that's how we approach everything."

Figure 1-3. Jon DeBaun has got his hands full.

adrenaline builds excitement. Excitement breeds the amazing performances that lead to legendary recordings! This book is about "adventure recording"—swerving down a country road with your head sticking out the moonroof. So let's buckle up our weird and find out where it takes us!

THE OPERATOR'S POINT OF VIEW

I'm not a technician, I'm not a programmer, and I can't even say I'm a super-duper expert at any of this stuff. I'm a designer of experience, a documentarian who captures fleeting moments of audio that I transform into an enduring legacy. I write from the perspective of an operator who is creating an audio postcard to the future. It was a shared experience when it was recorded, and you get a sense of those special moments every time it's played back.

So let's create music for all five senses. You should hear it, see it, smell it, feel it, and taste it. Music embraces all of life, and that's why this book has stories, tips, recipes, photos, advice, diagrams, exercises, illustrations, jokes, and pure fabrications. Because making music *is* all that and so much more . . .

A BANQUET FOR THE SENSES

When recording, how about making your session the most dramatic, sensual experience you

Jack Joseph Puig on Hitting Five Senses

"Keep in mind, we only appeal to one sense. We don't have the visual, like the film people. You can't smell a mix. You can't really touch it. Maybe you can feel it a little bit, because the speakers rattle—but the truth is, you are really appealing to just that one sense, the auditory sense. Somehow I need to make you feel that you can see it. That you can visualize what the singer is talking about. Visualize the person or the place or the mood he is singing about. Somehow I need to hit your five senses, even though I'm inputting into only one. And that is the magic! That is what we call the X factor. And any engineer worth their salt will tell you, 'I don't know how I do it.' Because if they did, they would hit them all every single time."

Figure 1-4. Scott VanFossen from Fighting Zero gearing up for overdubs.

MAYBE IT WASN'T SUCH A GOOD IDEA

Ed Stasium on Not Knowing Shit

"I still have no idea what I'm doing. I never studied theory or equalization or any of that bullshit. I just turn the knobs till I think it sounds good. I just tell the band to change the kick-drum part until it feels right—that's all I do. It's just my opinion. You don't need to go to school to do this."

Figure 1-5. Well, it was worth a shot.

Matt Wallace on Art and Emotion

"A true star has a unique perspective on life. They are ideally someone who has been to the edge, locked down, and has come back to tell you about it. These are the real artists. My whole thing is this: The only reason we have music, poetry, dancing, sculpting, painting is to convey emotions. That's it. If we could just connect from one brain directly to another brain and send emotion, we'd do it. But we need art to convey emotions. Emotions that are difficult or deep or resonant or challenging to discuss. Men in general have a hard time with emotions. So if you wrap it up in this thing called music, it's easier for men to talk about what they normally couldn't talk about.

"I make a joke about it every time I go to a movie with my wife, Melodie. We'll be watching a scene and the music starts coming up, and it's that 'heartstring' kind of music. The reason for that music is to cue the guy to reach his arm around his date and say, 'Oh, honey, I'm so sorry.' Because some of us guys are a bit thick. We don't live in the emotional arena as comfortably as women do, so it takes music, poetry, dance, sculpture to bring us to a place where we are able to feel, 'That's heavy, that's something.' Every time I talk to a band I tell them, 'The only thing we are doing here is peddling emotion.' Because we are going to make people feel angry, start a fight, want to sit down and cry, call up someone that they love, or talk about somebody who died. Everything we do is just to access these moments. Every great song connects with people. It might start off as a very personal song, but if it's done right, it's a song that becomes universal. The listener can hear him or herself in it. And when you can do that, that's when you have great art."

Figure 1-6. Hans Zimmer in his Remote Control studio headquarters.

Hans Zimmer on Cajoling Equipment

"Our job on a daily basis is to do something new and to do something that nobody has heard before. I mean, that's really the mandate here. So you start with a blank page every day. It's not like other jobs where you get really good at making rivets. There's no repetitive motion involved. You are always presented with a new problem. What you do get better at is knowing your instinct or your sense on how to cajole the equipment to achieve that sound in your head."

can conjure up? This is part of living life as an artist and living art as a 24/7 lifestyle. Be an artist in what you do, what you wear, how you speak, what you say, the food you eat, your surroundings, and especially what you do with music. Be sure you know *why* you are doing this. Are you recording music so you can be rich and famous? If so, then maybe you should reassess your choice, because recording music should be done for the love of recording. This is not to say that no one makes money. In fact, a few even get rich—very rich. But they started with great ideas and a desire to create.

UNBUCKLE UP

Mixed in with the nuttiness that is *Recording Unhinged* (this book, duh) you might find a phrase or illustration that will send you straight down the rabbit hole. The danger is, you may never find your way back. Time to check your fire extinguishers.

Ross Robinson on Not Playing Nice

"Instruments on my sessions get a decent amount of abuse because I don't want them to be played safely or perfectly—pretty much ever."

Figure 1-7. Hot organ licks!

Julian Colbeck on Being Precious

"People have become terribly and ridiculously precious about a sound or a performance. 'Oh, I've lost it!' Well, who cares? Do it again. It doesn't matter. It's not like someone chopped your hand off. Are you that untalented that you can't do it again? You can probably do it better, definitely different, and that's good! When a computer crashes—and they will—don't just say, 'Oh, a disaster!' It's not! You need to learn to be brave enough to trust your ability. It's a much more exciting place to be."

Figure 1-8. Devil Meter.

THE IMPORTANCE OF NOT HAVING A PLAN

Budgeting time to let the unexpected happen is very important. You must figure out what you'll need to do to satisfy the song, but also be sure to leave elbow room for magic. Some brave folks front-load their project with experiments, mistakes, and fluff. But then, when the deadline looms, they launch into action and get it done on time. OK—that works too!

HAPPY ACCIDENTS

I want to keep every outtake, even if it has obvious mistakes. Why? Because the best you can hope for in any session is for divine intervention. You would be surprised at how much popular music is peppered with what were originally thought of as mistakes, later recognized as genius. A lyrical fumble, an electronic hiccup, a slip of a finger. Be open for it to happen and keep those outtakes!

Nick Launay on Momentum

"I think the best records are the ones that you enjoy making. Sometimes you'll spend eight months on those records that are a pain in the ass to make. Where band members are not getting along, or it musically wasn't that good and you tried to make it good. Whereas some of the most successful records I've made were records that were done in a week. And they were incredibly exhausting and a lot of energy was put out to make them. But the adrenaline and the fun of making them made the difference. 'Wow, here's another idea, and here's another idea,' and so on. Those are the records, to me, that are the best records. The ones that had a momentum to them."

Figure 1-9. Nick Launay.

Elliot Scheiner on Steely Dan's "Happy Accidents"

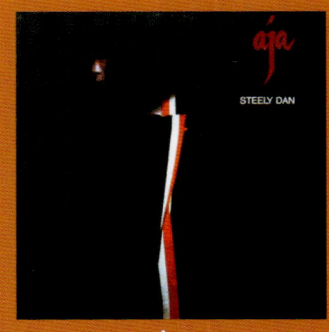

Figure 1-10. Steely Dan, *Aja*

"When I was mixing *Aja* for Steely Dan, on the very last cut on the album, 'Josie,' there's a guitar solo on there that I decided to put through a tape delay—an analog tape delay. And I had a mono Ampex 300 machine set up and we ran it through, but one of the two tubes started oscillating and it was causing the guitar to have this kind of warbley sound, and if you listen closely, you can really hear it in the mix. And they loved it! I asked, 'Are we OK? Because it sounds like we fucked up here.' But they said, 'No, no, this is really good. Let's keep this!' It's one of those 'happy accidents.'"

Brian Malouf on Whitney Houston

"Whitney Houston's vocal on 'I Will Always Love You' was a one-take performance. I was asked to do a mix with the vocal in the 'proper' perspective, but the board mix—done right after she sang it—became the record."

Figure 1-11. Whitney Houston.

ROSS ROBINSON ON ENCOURAGING MISTAKES

"Mistakes are God. Nonresistance. A mistake is the unknown coming through. The unknown that was never, ever, ever planned. A lot of times it will turn into something that you can build off of. It could be a beat, or a vocal, or anything. A breath. It's endless how many mistakes end up on finished records—I can't even begin to think. And I tell people as we're tracking, 'It might be tight in there. It's not there yet . . .' And I'll point at the drummer and say, 'I want you to make a mistake. I want you to fuck up! I want you to fuck up so much, and the more you fuck up, what am I going to feel?' and the band says, 'Happy!' And I'm like, 'Yup!' And it's crazy. It should be a golden rule of recording to encourage mistakes, because it creates a perfect take more times than not. The mistakes just go away. I encourage mistakes. And then they don't happen."

Ed Stasium on Gladys Knight's "Midnight Train" First Take

"Recording the vocal on the song 'Midnight Train to Georgia' by Gladys Knight & the Pips, I was just winging it. I was just starting out as an engineer, and Gladys said 'OK, I'm ready to do my vocals.' I remember putting up a Neumann U 67 through an LA-2A compressor, turning the knobs to some kind of setting, I don't even know what. I can't even remember what kind of console it was—maybe a Spectrasonics or a Quad Eight. I was just so new, I didn't even know what it was. Tony Camillo, the producer, said, 'OK, let's go,' and we recorded. It was one fantastic take, that's it. Luckily, my settings worked. We did wind up punching in four words in the vamp some weeks later, in the 'I got to go, I got to go' section . . . but really, that song was one pure performance."

Figure 1-12. Gladys Knight at Record Plant, New York.

ENGINEERING MARVELS

Figure 2-1.

2

ENGINEERS AND DEMOLITION EXPERTS

FEARLESS RECORDING!

Who says you have to record in a treated room with high-end condenser mics? Who says the compression should always follow the EQ? Who says ribbons are the only mics to use on horns? Who says vocals must be performed in front of a 60-year-old, $10,000 dangling German tube mic? Whoever says it—they are wrong! There are no rules! There are only *suggestions*, and many of them should be taken as someone else's idea—not yours. Go ahead and make shit up. Create a new instrument out of kitchen items. Use a mobile phone as a delay device. Record the guitar in a helicopter over the city of Prague. In fact, the sky is NOT the limit! There are no limits!

ELLIOT SCHEINER ON STUPID IDEAS THAT WORK

"I started in the late '60s, when everyone was a staff engineer and people would get hired because maybe you could bring more business to

Figure 2-2. Per Kristian Sundet records bass guitar in a torpedo tube—yes, in a real submarine.

one studio over another. In those days it was about 'What can you do differently than the guy who is in the next room, or the guy who is in another studio?' I tended to want to do things differently to make people confused: 'Does this really work?' 'Is what you are doing OK?' I got a call to do an album with an Australian artist. The producer was Noel Stookey from Peter, Paul and Mary. We were working in a small, boxy room at A&R recording, maybe 30 by 30 feet, and the ceilings were maybe 24 feet high. And the leakage in this room was problematic. Nothing ever really worked out great in there—you'd always go, 'Oh, God, I wish I could have done this somewhere else.'

"So I thought maybe I would help that situation and wow the band at the same time by doing something really weird. I wanted to do something that the band would come in there and go, 'What the fuck?' So I put up two sets of scaffolding along one side of the room, one for the guitar, and one for the bass. The only thing on the floor was drums and some keyboards. I tell ya, when everyone walked in, they looked at this thing and freaked out. There were ladders so that you could get up there and adjust their amps and stuff. Nobody had ever seen anything like this. Engineers would come in that were in neighboring rooms in the studio and go, 'What the fuck are you doing?' The perception was that this was something so different that it had to work, it was so cool! It really didn't; it didn't do anything differently, y'know. I still had as much leakage, but the idea was that these artists were so impressed with something as stupid as this, that if they had the chance, they would probably come back and we would do the next record as well!"

LET IT BLEED!

I like to have the musicians all playing simultaneously in the same space. The unspoken com-

Ed Stasium on Soul Asylum's Ampeg Catching Fire

"I had an Ampeg Jet amp that I bought for 60 bucks from some old guy on 48th Street on Manhattan's Music Row. I brought it into the Soul Asylum *Hangtime* sessions at Right Track, where we were doing a song called 'Clean' and David Pirner was playing guitar. All the sudden, smoke started pouring out of the Jet. Flames shot out of it. We called the session the 'Sacrifice of the Ampeg,' because whatever it was that we recorded was kept on the record. It made one really weird fantastic fucking noise."

Figure 2-3. Ed Stasium with Soul Asylum.

Geoff Emerick on Being a Troublemaker with the Beatles

"Being a young mastering engineer opened my ears to the fact that there were different sounds coming from American records than there were coming out of EMI studios. So knowing that, I was striving for new sounds if I could possibly get them. I was 19 when I started to do *Revolver* by the Beatles at EMI's Studios at Abbey Road. 'Tomorrow Never Knows' was the first track we recorded, and they presented to me the fact that they were fed up with the old drum sound and the old bass sound and guitar sound, because they'd heard American records too. But there was a protocol at EMI. For instance, you weren't allowed to put the bass drum mic closer than eighteen inches away from the bass drum, because the air pressure would damage the diaphragm in the microphone. I knew there was this more 'up in your face' funk that you could get, so I think that was the first time we took the front skin off the bass drum and stuffed it with some stuff and put the skin back on and moved the mic closer, which I got into trouble for. Everything went through a 660 Fairchild limiter, and I was overdriving it, but that whole thing gave us the new up-front Ringo drum sound."

Figure 2-4. Abbey Road Studios.

Figure 2-5. Garth Richardson gives former assistant Joe Barresi a special haircut.

Garth Richardson on Recording Rage Without Handcuffs

"When I recorded the debut Rage Against the Machine album, we set up the session like a live show. I didn't want the band to walk into the studio with handcuffs on, so we actually put everything through a concert PA system, and everyone tracked with no headphones on. It was so loud that the studio manager at Sound City could not talk on the phone. Studio B had to stop working unless they were doing a song that was in the same key as us. Erad [Wilk]'s drums were set up behind the PA. Tim [Commerford] and Tom [Morello]'s amps were isolated in the back room. On one song, 'Settle for Nothing,' Tom was not happy with his solo, so we overdubbed it. You can still hear the original live solo track bleeding through the drum mics. We got lucky—it sounds like an echo! We even kept Zack [de la Rocha]'s live vocals up to the point where he lost his voice. When we retracked it, we had him in the control room holding an SM57 with the monitors turned all the way up, blasting."

munication between players is uncanny and difficult to recreate if everyone is separated by glass or being recorded at different times. Brave simultaneous recording is also an engineer's nightmare. Bleed between instruments is virtually impossible to correct after the fact. With electrified guitars and bass, there are a few speaker isolation techniques that will help keep sounds separate while still maintaining the live feeling in a performance.

I also like to have the musicians stand in the room with the drums, with the guitar and bass heads driving cabinets in an adjacent room where they are miked up and being recorded. The lead singer can sing in the room with the group, giving guide cues for the band to follow. It doesn't matter that the vocals are sung out in the room with sensitive overhead drum mics—rock music is usually loud enough to mask the bleed.

And so what anyway? To get the best "live" feel of a recording while maintaining some separation of instruments, you can also set up a small PA in the tracking room with the players, feeding the isolated cabinet's return into the PA, where it is set up to play in the open room. That way, the musicians can shed the headphones. And the idea of in-ear monitors makes the live performance in the studio easy as cake. Just make sure your band knows the song!

Here's a crazy move: Put the recording equipment in the same room as the musicians. This can actually help in

Figure 2-6. Marius free-ballin' the Seigmen sessions.

Matt Wallace on Paul Westerberg's Toilet Session

"At RPM Studios in New York, we were doing Paul Westerberg's first solo record, and it was also his first sober record, so he was a little nervous. He had this band all ready to go, and we were working with Susan Rogers, who is an excellent engineer. So he had the band, he had the engineer—I thought, *Oh great! Let's do this!* So on the first day, we get a song done. Fantastic! At the end of the day, we said, 'See you tomorrow,' and we all took off. I show up the next day and I just see Paul, and I ask, 'Where is everybody?' Well, he fired everybody.

Well, I'm thinking this session is doomed. But I was smart enough to have packed my trunk of musical and audio crap. Paul went into the toilet and sat with a twelve-string acoustic guitar and was working out a song. I went in there to pee and thought, *Hmmm, that's an interesting song,* so I dug around in my trunk and pulled out a Fostex F-16 multitrack cassette deck that I brought along for the trip. I grabbed a microphone and a cassette tape, walked back into the can, set it up quickly, and started recording. That ended up being a song on the record called 'Even Here We Are.' It was so noisy, you can hear the hum of the air conditioner. I had to filter that out, but it was one of those things where it was instantly 'Let's do this.' It wasn't 'Let's do this the right way.' You don't want to lose the moment. You have to shoot from the hip and trust your instincts."

many ways. Communication between the producer/engineer and the musicians is immediate. The producer/engineer will know exactly what the instruments sound like, instead of guessing through the glass. The musicians feel less isolated. Of course, the producer/engineer will need headphones that they are familiar with and that they trust, and the recording crew needs to play a sample recording back in the air to know what they are getting. I have used this technique regularly for the last 15 years.

One drawback is that the band can hear you talking shit about them. My engineers and I are proficient at passing notes back and forth during tracking, crinkling up the notes, and throwing them in the trash after reading them. Luckily, musicians are not likely to go through the trash bin.

I suggest in any new setup, when an important session is coming up, that the recording crew test the system by recording a "surrogate" band—one with the same instrumentation as the band that will be coming in. I did this with a Smashing Pumpkins session I engineered for Rick Rubin at Sound City Studios. I had a friend's band come into the studio the night before to test my whole setup, so I could dial in sounds without pressure. When Billy Corgan and the band showed up the next day, we had the session up and recording within two hours, and it sounded amazing. Ya-haaa!

Figure 2-7. Courtney Love knows how to make noise.

Linda Perry on Courtney Love and the 666

"I have this old funky mic called an EV 666; it is an old Beatles microphone. I love it for when I'm getting down and dirty and it's got to be gritty. I've put that on Courtney Love, because I'm looking for punk-rock, edgy, raw sound. I'll put it on a stand, but if somebody wants to hold it, go for it! I don't care about all the noises everybody else is so concerned about. Like, 'Don't you hear that snare bleed?' I'm like, 'Yeah.' But I'm not going to hear it when there's all the drums, all the guitars, all the bass! I'm not going to hear the air-conditioning click on!"

Larry Crane on Knowing When It's Done

"When is a song finished? Well, the cynical part of me says, 'When the money's gone.' But the creative side of me says it is when it feels right to everyone. We kind of make a laundry list of everything that needs to be done at the beginning and tick away at it throughout the project. And when you are done with that, you sit back and listen and see where you are at and make a decision on what is left to do."

COMMITMENT, BE BRAVE

In a world of endless tracks and playlists, committing to sounds and takes has fallen out of favor. I challenge you to change that attitude. You will find yourself sleeping much better at night knowing you've finished something. You can file it in your head as "done." This allows you to move on to the next challenge, the next task. Don't be afraid to finish. It is something you may need to practice to really get right. And when you commit to that track being finished, or that session being completed, or that mix being finalized, get rid of all the extras, the drafts, the "maybe" takes. Back them up and get them out of the session, or at least out of the edit window. Consolidate your tracks. Be brave—you can do it.

RAINBOWS, UNICORNS, AND THE AGONY OF DEFEAT

Not all days will be rainbows and unicorns. Don't beat yourself up over it. There will be times when nothing works, when your best idea ever turns out to be a dud. You might burn a day on some detour only to realize it's a dead end.

I once had a plan to throw a guitar off a cliff during a solo, recording it as it tumbled down the hill. Fantastically brilliant, right? The thought was that it would sound like a shrieking monster as it crashed to the bottom. The project was the Machines of Loving Grace album *Gilt*.

We prepared a sacrificial guitar for the event, painting it vividly and drilling a hole through it so we could attach a rope. Then we dragged a Marshall amp with a very, very long extension cord to the

---

Geoff Emerick on Robin Trower's Piano Resonator

"When I was recording Robin Trower's *Bridge of Sighs* album, I thought of a sound, which was some sort of shimmer on the top of the guitar notes—I had the idea in my head for just the right effect. All of Robin's amps were set up in the studio, so what I decided to do was put an additional Marshall on its back looking up underneath the piano's keyboard, hoping to excite the strings in the piano. But then I didn't realize that it was exciting the strings that we didn't want. So, yeah, it was ugly. But had I realized that we could play the chord and press the notes on the sustain pedal and then only those strings that we wanted would resonate—well, it might have worked. But it was getting a bit labored by that time, and we weren't really getting it to do what I wanted it to do, so we moved on to something else. Now, if I had pursued it, I might have gotten something out of that idea."

Figure 2-8. Sylvia with Geoff Emerick.

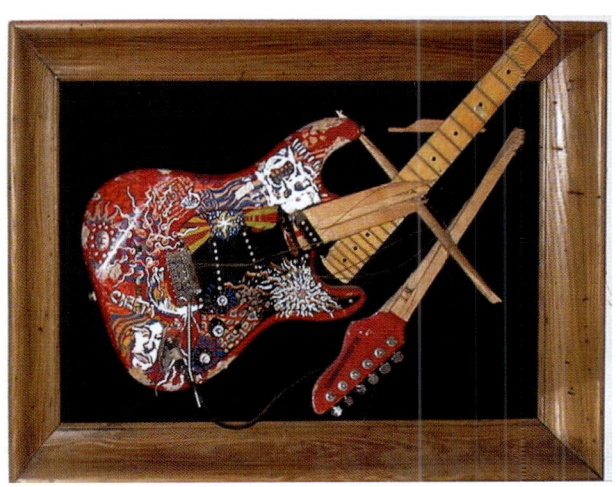

Figure 2-9. Sacrificial guitar trophy, after being thrown off a cliff.

top of a cliff overlooking the ocean in Malibu, at a recording studio named Indigo Ranch. I set up a stereo pair of mics facing over the cliff and connected the guitar to the amp with a 100-foot instrument cable. The guitar player, Tom Coffeen, stood at the edge of the cliff and turned up the volume, causing the feedback to swell. At the pinnacle of the feedback noise, Tom threw the guitar over the cliff and the crashing resounded through the canyon down to the ocean.

Wow! We retrieved the guitar by dragging it back up the hill with the rope. The neck had snapped off and was being held on with strings. It was fantastic! We cheered and high-fived and went inside to place the "solo" into a track on the album. Unfortunately, the sounds that had been recorded just did not fit into the song. We tried using the cliff-toss cacophony as a segue between songs. Still it did not feel right. Wherever we tried to put it, the cliff feedback just sounded forced. So we abandoned the idea. A full day of recording was lost, but the memory of that event will *never* be forgotten. One of the best rock-'n'-roll moments ever.

TECHNICALLY, IT'S ALL NOISE

"Love the one you're with . . . " As luck would have it, wherever you are is probably good enough to record. I've recorded in moving vehicles, hotel rooms, bedrooms, kitchens, garages, theaters, tents, bathrooms, coffee shops, churches, barns, bedrooms, clubs, chambers, basements, outside in the yard, in the pool—doesn't matter. Your recording environment becomes an effect, an additional instrument. Use the space. The ambience is real and alive!

THE RIGHT GEAR MAKES IT EASY

I admit it, I am a gear snob. But not in the way you would think. I don't mind using inexpensive mics, but what those mics go into makes all the difference in the world! So my mic pres are killer and not cheap. Beyond that, I feel it is important to have some essential studio tools on hand—tube and optical compression, quality EQ. And fun, weird shit to get lost in. I've

George Drakoulias on Maria McKee's Fake Live Record

"Directly across the alley from Hollywood Sound Recorders was a club, so we made a fake live record with Maria McKee. We put Maria in the bathroom to sing her vocals, and the sound of the club, with people just milling about, kind of drifted into the open bathroom window from the alleyway. It sounded far away, and it felt really good. It sounded like she was working in a club and no one was paying attention to her, which was part of the joke of the song—it was the cover of 'Wichita Lineman.' Benmont Tench on piano and Maria was in the bathroom."

selected a few items to include in this book, though there is far too much to share in this single volume. So consider this an introduction. Remember, certain gear will make it easier to achieve that sound in your ear's eye.

The evolution of the recording studio is fascinating, starting from Edison's phonograph with the input sound-collecting horn, needle scratching indentations into the wax, playing back the sound through the horn—so simple. Eventually, that grew and improved into the vast fields of knobs and switches called recording consoles. Then it was all reduced into illuminated screens with interactive images of these beast boxes. Amazing, all of it. Oh, and then you have Jack White and T-Bone Burnett going right back to the beginning!

As far as initial tracking of instruments in the analog age, arguably the most popular consoles have been made by Neve, API, Trident, and Solid State Logic (SSL). There are other great "honorable mentions," such as Helios, EMI, Focusrite, Sound Techniques, Quad Eight, Crystal, Calrec, Harrison, MCI, and the legendary Universal Audio consoles. But that is for a whole other book! We'll concentrate on only the most popular analog monsters here.

Figure 2-11. Evolution of recording.

Matt Wallace on Blindfold EQ

"If I were King of the Universe, consoles would have no indication of frequency near the EQ knobs, because when you show the frequency, then most people EQ by eye. I was guilty of this. I would work on my SSL, and I'd say, 'Oh, I really like the 2 K for the blah blah,' and I'd go right for that thing. This is what you need to do instead: Close your eyes, sweep the EQ all the way up and find the frequency you like and how much of it you like, and leave it at that. You might look and go, 'Oh my God! 8 dB at 3 K! You're not supposed to do that!' But seriously, you should have no idea of what frequency you are boosting or cutting. It doesn't matter what the number is. It matters how it sounds and how it feels. That's it!"

Figure 2-10. Matt Wallace.

THE ERA OF GIANTS

Why is it important to know about these original large-format multichannel analog behemoths? Because so many great albums were recorded on them and they have a sound that is instantly recognizable once you know them. And because so many modern software and equipment makers try to emulate the sound of these original circuits. And if you are lucky enough to record on the originals, you may find yourself ahead of the game.

Figure 2-12.

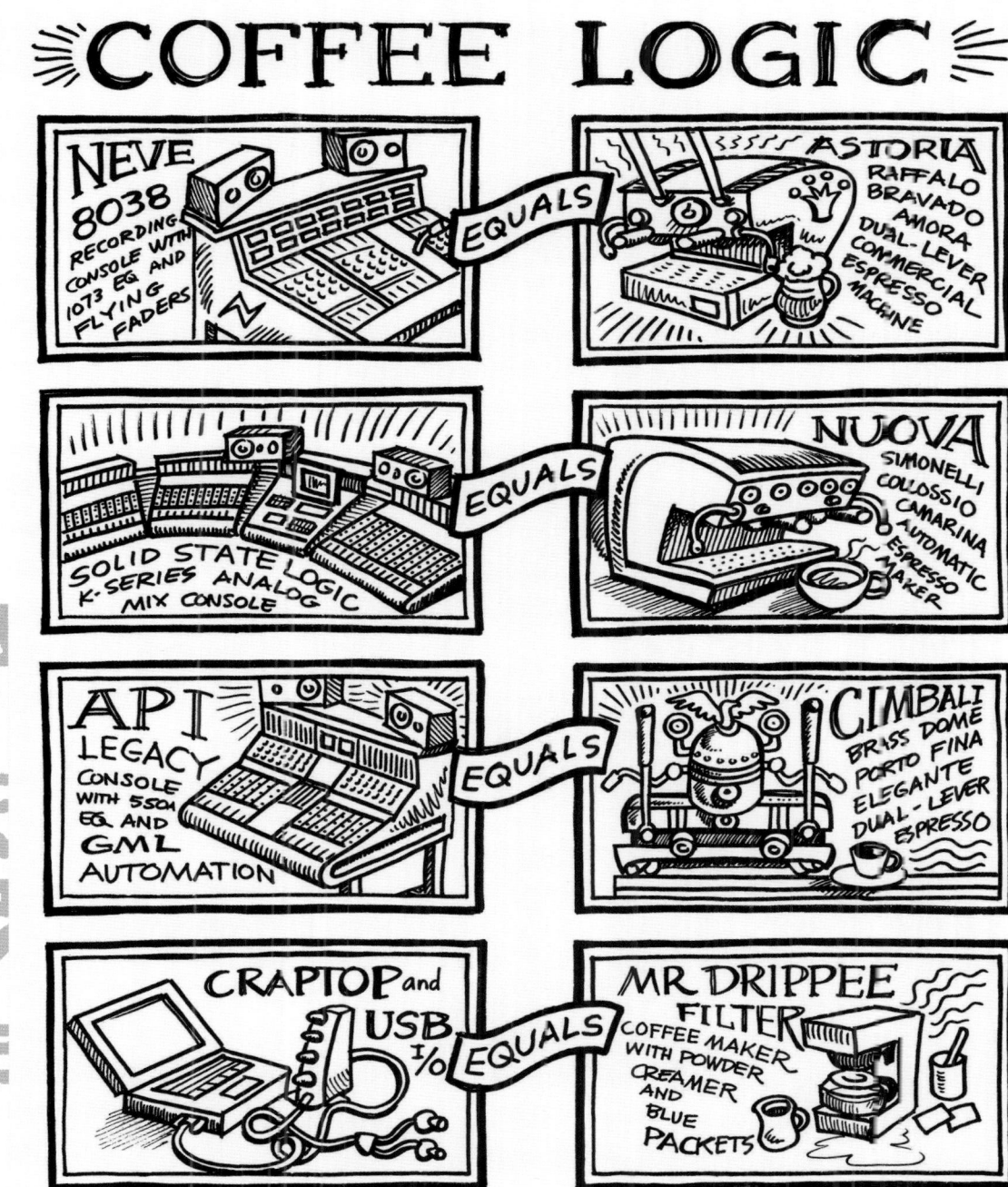

NEVE

Arguably the most copied mic pre/EQ circuits are from the original Neve consoles, built between 1970 and 1980. This is when an Englishman named Rupert Neve owned and ran the company and technicians like Geoff Tanner implemented Rupert's concepts through the design and construction of the consoles. These larger consoles are often called the "80 series," with model numbers such as 8028, 8038, and 8048, generally featuring a split-console configuration and bussing matrix of 16 switches for routing to the recorders.

Figure 2-13. My best friend, the Neve 8038 console.

These earlier Neves were generally custom-built modular desks, with the company offering several types of class-A mic pre / EQ modules to suit your needs. The most popular were the 1073 and 1084 EQ modules. Smaller 80-series configurations also existed in the early '70s, with 8016 console models custom built with four or eight busses. These and the even smaller BCM-10 broadcast consoles were also offered with the class-A style mic pre/EQ modules. In the late '70s, the consoles were more standardized, with the popular 8068 and

Figure 2-14. Dusty Brill, drummer from Good Charlotte.

Eric Valentine on Using His Unpopular Neve to Make Hits

"Before I built my own console, I pretty much used Neve consoles exclusively. My first 'real' desk was a Neve 8038 console that I bought from Ocean Way. It had 32 1081 mic pre/EQs in it, with a 24-channel monitor section—an incredible tracking console. It just wasn't quite flexible or powerful enough for mixing for me. So I decided to sell it and got a 56-channel Neve 8128, which is not a very popular Neve console. Many people actually despise that model—I personally think it is an incredible console. I made probably all my best records on that desk! All the Good Charlotte recordings were done on that Neve; also Queens of the Stone Age, Smash Mouth. It was definitely not the vintage class-A thing, but it had a sound to it and its own personality. I got so used to the way that thing colored the sound that I still keep a little bit of it around at times when I know I need this particular kind of high end. That was the only console that did it. So I still have some of the 8128's EQs around."

8078 consoles featuring class A/B electronics and the mic pre separate from the EQ module.

When Rupert left the company in the early '80s, Neve lost a bit of its panache, but came back in the '90s with the innovative and smartly marketed "Neve V Series" consoles. Today, led by the man himself, Rupert Neve Designs builds analog consoles that scratch the same recording itch as his original namesake consoles.

TRIDENT

At the beginning of the '70s, Trident Studios in London's Soho district had a crack tech team designing and building their own custom consoles, with the legendary Trident A-Range console being one of the most desirable, followed by the hugely popular 80-B and 80-C. Malcolm Toft started the trend by working together with designer Barry Porter to create the custom A-Range consoles for Trident. The distinctive sound of the A-Range Trident can be heard on the Beatles' "White Album," Queen's *Sheer Heart Attack*, Marc Bolan and T. Rex's greatest hits, David Bowie's *The Rise and Fall of Ziggy Stardust*, and even Devo's *Duty Now for the Future*. L.A.'s Cherokee Studios was one of the first in the States to discover the Trident sound, and the engineers there ordered three A-Range consoles to install into their Fairfax Avenue facility.

The next generation was the Trident Series 80 con-

Figure 2-15. Joe Satriani.

Figure 2-16. Trident Series 80 console.

sole, reasonably priced compared with Neves, with some slight improvements. John Oram joined the team in 1974 and was essential in the improvement of the Series 80 Tridents. The company was able to build its consoles at a lower cost because these consoles did not have separate enclosed modules for each EQ, bussing, and auxiliary section. Instead, Trident mounted the electronics for each channel on one neat channel card, which lifted out easily for maintenance. These consoles were extremely popular in the late '70s and '80s, becoming a mainstay in many commercial studios.

The first time I worked on a Trident console was in Studio B at Hyde Street in San Francisco on their sweet little 80-B. That was with Joe Satriani on the *Surfing with the Alien* album, with John Cuniberti at the controls. I was just starting out. I eventually bought one of those Tridents. Loved it!

API

Across the pond in the '70s, the American company API was building custom consoles that rivaled those of its English competitors. These were also modular consoles, like the 80-series Neves, and were stable and sweet-sounding, with the flagship 550 proportional Q equalizers. The "penthouse" section under the meter bridge generally held the API console's EQ modules. Custom builders, including Frank De Medio, used API components in their console creations.

Today, Frank's consoles are some of the most famous of the vintage APIs around. Originally, there were no other companies building the now well-known 500-series modules, but years later, this simple design led several other audio designers to use the same format, allowing the other manufacturer's EQ modules to be used in many API consoles.

Figure 2-17. Custom-built API console.

Figure 2-18. Paul Wolff.

In the same way, studios building their own custom consoles could use the popular API 550, 550A, 554, and 560 EQ modules by installing the racks into their consoles to accommodate API 500-series modules. This is the standard that is used in today's 500-series racks.

The production of API consoles slowed in the early '80s, and the company was all but done by 1985, with the original crew nearly gone. However, an audiophile named Paul Wolff, recognizing the value of the API sound, revived the API brand with the "Legacy" series in the '90s, and cleverly marketed the EQs in their own "lunch boxes" so anyone could easily use API gear in their studio. Paul no longer owns the company, but the current owners of API enjoy success with their new generation of consoles based on the original designs.

Figure 2-19. Prince loves those API De Medios.

Susan Rogers on Jumping into Prince's API De Medio

"At the studio in the basement of Prince's new home in Chanhassen, we had just taken delivery of and installed a custom Frank De Medio–designed API console. I call it an API, but it was really a De Medio console with API EQs. It was patterned after the one in Sunset Sound Studio 3 that Prince loved so much. We just finished connecting the last wire, right? Prince, in waiting, had naturally written a lot of songs, but uncharacteristically for Prince, there was a song that Lisa Coleman had written that Prince wanted to record immediately.

"It was beautiful, beautiful, beautiful. Lisa wrote the music, and Prince wrote the lyrics, and he called it 'Power Fantastic.' So, Lisa was upstairs on the piano, Bobby Z. was on drums in the one booth that we had, Wendy Melvoin was on guitar direct, Brown Mark was on bass direct—I didn't have enough headphones for everybody, so Prince had to do vocals in the control room with me! It's Minneapolis in the middle of the night; it's not like I'm going to run next door to get another set of headphones! He had the last set of headphones so he could hear the track. Normally, he would never work under these conditions, but he had me put the mic up in the corner of the room, and he turned to face the corner while he did his vocals. He had his back to me, so he could have that privacy that he needed. So I recorded his voice, listening to nothing, because I am recording on the API and the monitors are turned down. It was great hearing his beautiful voice all alone."

Ed Stasium on Spitting on the Fire

"I used to have a great deal of volume in the control room while recording. That was until a Yamaha NS-10 woofer caught fire and sparks shot out of it during a Fetchin Bones session at American Recorders in Los Angeles. The speaker seriously caught fire, and we had to put it out. Actually, I spit on it to put it out! Well, the studio had Richie Podolor's Trident A Series console. I'm not going to put the fire out by pouring water on it! I had to think quick! Now, I listen at very minimal levels unless I'm tracking drums. Actually, these days I mix on powered 'Partner' speakers by Advent. I've had them since the early '90s. Garth Richardson uses them also. They were a computer speaker design that was pretty decent for the price."

SSL

Back in England, the late '70s were overrun with a new generation of large-format consoles, with the popular Solid State Logic consoles taking over the landscape. The many great SSL improvements include standard in-line formats and on-board computer mixing with a recall feature that allowed the user to take a snapshot of where the knobs and faders were set at any time. You could leave the mix, change the settings, and, with the Total Recall, be able to bring back all the original settings.

This was a huge advance in studio mixing of music. Thousands of these E, G, J, and K consoles were sold worldwide and became the "workhorse" desk for commercial studios. However, many people felt the SSLs didn't have the bold sound of the Neve in the mic pre-amps. To get around this, professional sessions were often recorded on the Neve, Trident, or API, and then carried over and mixed on an SSL in a different studio. In the '90s, this was how many records were made.

In 2006, Peter Gabriel—the amazing musician who is also a daring and innovative businessman—purchased the SSL Company with his broadcast entrepreneur partner, Dave Engelke. They have created a new generation of consoles, plug-ins, and rack-mounted audio processors that embrace the digital trends in recording. One of their newer designs is the Duality, which combines the functions of a digital controller with a "large-format" yet virtual analog console. This is some *goooood* stuff!

Figure 2-20. Peter Gabriel.

OK, NOW WHAT?

So, get your hands on some of these classic mic pres and EQs if you can—they can be racked-up originals, or they can be knockoffs. Lots of companies are making reproductions. You can use the digital versions, too, but plug-ins probably won't react the same way the originals do. Especially if you try to abuse them the way you can abuse a rack unit. And you might just want to push the limits. Don't be afraid of going too far!

CREAMING THE MIC PRES

Figure 2-21. The glorious Solid State Logic J-9080 console.

Driving the mic pre is typically a no-no, because it can potentially add an unpleasant distortion on the track you're recording. But certain mic pres sound more exciting the harder you

Al Schmitt on Early Equalizers

"When I started, we didn't have any equalizers. We had only one equalizer—a 'Cinema' equalizer. And if you used it, it equalized everything, because we were recording mono! So you couldn't just put it on the bass or put it on a vocal or anything, so we rarely ever used it."

Figure 2-22. Al Schmitt.

Figure 2-23. Neve 1073 Modules.

drive them, so dare to test the limits on the input level of a mic pre. Use the distortion as an effect. Putting a rock-style guitar through a Neve mic pre on the lowest setting will sound good, with lots of dynamics and clarity of tone. But if you want more excitement, try driving the sound with the mic pre instead of increasing the gain on the guitar amp . . . you will be surprised at the "natural" compression the signal undergoes. Try the same idea with a vocal microphone. You will not hurt the mic or the circuitry of the mic pre; however, you might think you've maxed out your mic pre, while in fact what you are hearing is the clipping of digital distortion from hitting the recorder too hard. The digital clipping sound is not nearly as pleasing as the aggressive use of mic input, so depending on what you are going for, you will be able to get the right kind of nastiness you want by knowing where the nastiness is coming from!

UNTAMED DOUBLE COMPRESSION

Another unconventional secret weapon is the use of some very particular and somewhat strange compressors chained together. Don't be shy—just slam audio through them! I personally will use more than one compression unit patched in succession to squeeze every last drop out of that sound. I want to hear the pulse coming off the jugular of the singer when I'm recording vocals!

Hard double compression will give you that close intimacy while knocking back the loud stuff. It allows the singers to really do their thing without having to direct them to back off the mic in loud sections. If spontaneity is king, then technicality is queen. Be careful with the shrillness that sometimes occurs when you crush audio. The lower-grade compressors will thin out on the louder signals and sometimes distort easily. That is why the best engineers and

Figure 2-24. Clare Pproduct breathes fire.

Linda Perry on Double Compression

"I love double compression on vocals. There are a lot of people that don't have air, and I feel like it helps push it out. Also, having two compressors in series brings the voice so close in the singer's headphones that they stop over-singing, because they can hear every little lip smack and every little detail in their voice. It makes them start singing from a warmer, more 'meat-and-potatoes' place."

producers use very specific equipment. Again, the right equipment makes the job easy! Here are some of the best-loved and most closely held secrets of compression: the broadcast compressors!

CREEPY OLD GEAR

Incredibly, *waaaay* before the recording studio boom of the '70s with all the new music-related gear, the broadcast world had been in a boom time—since the '50s, in fact. Dozens of strange and wonderful American companies designed tube-based compressors for radio stations and radio production. Manufacturers like Gates, Collins, Federal, General, CBS, Valley People, Orban, Symetrix, RCA, and Western Electric made spectacular units for compression, gating, limiting, and processing for the broadcast world.

I discovered these broadcast audio treasures in the early '90s after working with Los Angeles producer Jimmy Boyle. He brought several pieces into the studio that I had never seen before. They had industrial gray fronts and old art deco meters and outdated anything I had in my rack. I immediately set out to find my own crazy vintage broadcast compressors. On a trip to Michigan, I went through the local phone book and cold-called the radio stations in the area. The broadcast industry was going through a digital revolution at that time, making the old analog equipment obsolete.

Figure 2-25. Extra creepy old stuff.

Lucky me! These classic compressors were being sold for dirt cheap! I connected with a broadcast salvage/liquidation company out of Toledo, Ohio, and had several shipments sent to me of RCA, Gates, and Collins beauties, till my whole living room was filled with stacks of old gear.

And they were creepy-looking devices. Many had not been racked for twenty years, just thrown in a pile in a back room of some radio station. Dirt and dust and broken tubes and leads. Poor things. I gave them a new life in recording studios of Los Angeles. Broadcast compressors are often slow and heavy and have a mid-frequency bump, which gives it an old-timey flavor. Here are a few treasures to discover.

STA-LEVEL

The Gates company made equipment for radio stations starting in the mid-'50s, with the popular SA-39B, Level Devil, and Sta-Level compressors. It also made broadcast and remote mixers and amplifiers. The Sta-Level compressors were strong, slow, and colored the signal with a mid-range boost.

Figure 2-26. Johnny Cash during the recording of *Unchained*.

Figure 2-27. Gates Sta-Level compressors.

I used '50s-era Sta-Level compressors on Johnny Cash's *Unchained* album in 1996. It gave his voice a warm, nostalgic feeling. Recognizing the value in the use of these vintage Gates compressors in music recording, Phil Moore started Retro Electronics and reverse-engineered his Sta-Level compressor to be as close as possible to the original, which had not been built since 1971. The Retro Sta-Level also uses the 6386 vacuum tube, the same tube found in the Fairchild compressors. This gives even the reproduction compressor that warm but muscular, soft and present sound that is so desired in recording.

Eric Valentine on Frickin' Cool Blown-Out Vocals

"Sometimes on a vocal, I'll have a clean track, then take a mult of that and put it through an old Gates Sta-Level, and then just turn both the knobs all the way up! At the same time turning the gain on the channel strip all the way up! You get this completely blown-out, distorted, super-fuzzy thing, and you just mix a tiny bit in with the clean track."

Figure 2-28. Eric Valentine at Barefoot Sound on his Undertone console.

Figure 2-29. Collins 26U limiting amplifier.

Figure 2-30. RCA BA-6A.

COLLINS

The 26U is another handsome rack-mount compressor from the '60s and '70s. The Collins Radio Company made broadcast equipment that rivaled Gates and RCA, two other top-notch American companies. The 26U units also used the 6386 vacuum tube, and functioned with the same muscular ferocity as the Gates and Fairchild, with a slightly different audio character. Slow, warm, and strong, these compressors are a favorite of mine for bass recordings.

RCA

The standard of strength in the radio world, RCA built incredible broadcast mixers, amps, and compressors, including the highly desirable BA-6A compressor/limiter. I discovered how wonderful these units are on vocals and acoustic guitar, bringing out every nuance while adding their own special vintage color.

Michael Beinhorn on the RCA BA-6A Compressor

Figure 2-31. Michael Beinhorn.

"My favorite vintage tube compressor is the RCA BA-6A. I love it because it's atypical of most older tube compressors. It's bold, hype-y, relatively fast for an older tube compressor, helps maintain tonal distinction from other sounds (instead of gluing them together), and makes nearly any program that's put through it absolutely enormous. It doesn't work on everything, but when it does, it's delightful. If I use it in a vocal chain, I'll usually put it at the end, since it doesn't have timing controls (which makes it not as flexible for level stabilization)."

THE MISH-MASHERS

The world is full of exciting unexplored compression. Dig around in the back of garages, radio station storerooms, and technician's basements and maybe you'll find something with potential. Such devices are fun and sometimes slightly dangerous. For instance, I found this weird thing the back of some fellow's garage in Glendale, California. He had worked in a radio station at some point in his past and had been carrying around this ancient Western Electric 1217 limiter for decades. The front panel was missing, exposing the guts from the enormous broadcast unit, wires hanging out everywhere. I carefully plugged in the old compressor anyway and ran a guitar signal through it. The sound was obnoxious and alive. The Western Electric, I'm sure, was not working properly, but what it added to the guitar sound with its special distortion was something completely unique! So instead of repairing it, I've left it exactly the same. It's just the right amount of broken. It's now nicknamed "the Army Man" because System Of A Down singer Serj Tankian added a small plastic army man into

Figure 2-32. Western Electric 1217 "Army Man."

Figure 2-33. Øystein Greni from Norway's Bigbang adjusts a knob.

Shelly Yakus on the "Oscar Mayer" Limiter

"After recording the Raspberries song 'Go All the Way,' we tried mixing it once or twice and the song just laid there. I remember producer Jimmy Ienner saying to me, 'I can't believe this, I was counting on this song. Now I don't even know if we can put it on the album.' So finally I said, 'You know, we got this limiter the other day called a Roger Mayer. Maybe we should try it.' This was a prototype unit that I'd tried earlier that week at Record Plant in New York. It was so violent-sounding, we started calling it an 'Oscar Mayer'—put music in one end, get hot dogs out the other! It was the first one that Roger Mayer built, and he sent it to us at the Record Plant for evaluation. All the parameters were Roger Mayer's parameters. So I said to Jimmy Ienner, 'You know, let me try this limiter on this mix.' So we took one of the mixes that we had and put it through this thing, and somehow the release time of that limiter was the same tempo as the song. Every time the song would let off for a moment, the limiter would let off in perfect time.

"It started to pump, and suddenly the song sounded like a Top 5 record. Jimmy Ienner turned to me and said, 'I don't understand. This song went from not even being on the album, to being maybe a number one record! How is that possible?' Well, unfortunately after that mix, Roy Cicala, owner of the Record Plant, sent it back to Roger Mayer and said, 'Look, the limiter is too crazy-sounding. I can't use it for anything.' And Roger changed the parameters. So when I went to use it again later on another song, it didn't sound the same. It didn't do that thing anymore ever again."

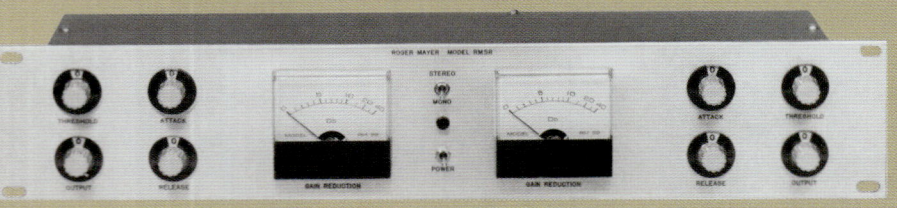

Figure 2-34. Roger Mayer RM58 limiter.

its exposed guts. I roll the Army Man into sessions and plug it in only when I know there are no children in the room.

Other great broadcast compressors include Spotmaster, CBS, Orban Optimod, CCA, Federal, General, and the list goes on. Don't be afraid to give them a shot; however, prices on Sta-Level and RCA units have gone through the roof. You might pay through the nose for the chance to play around with one channel of fun broadcast compression.

MY BESTEST FRIEND EVER

These are rare and fine birds. At $15,000–$30,000 per unit, Fairchild compressors are the most prized of all the vintage compressors, if you can even find one. I heard a rumor about someone tripping over one of these being used as a doorstop back in the '80s. I think it was Alan Dickson from Grandmaster Recorders in Hollywood who told the story, but then, he was always prone to tall tales! No matter. Run your mix through the stereo version of the Fairchild, the 670, and you know what the hoopla is all about. It makes the mix sound finished. Everything gets heard just sweetly and perfectly mooshed together. These were originally designed to be used for mono mastering lathes, with one channel processing the vertical movement of the cutter and the second channel processing the horizontal movement of the cutter, so the two sides of a 670 are never quite balanced. You can make up

Figure 2-35. Matt Wallace's dual CBS unit.

Matt Wallace on Maroon 5 Vocals Through the CBS Volumax

"The CBS Volumax compressors were made for FM radio to keep the signals from overmodulating. I bought this two-channel unit thinking I could use this as a stereo pair, but both sides sound really different. Because of this, I have one side named 'Simmering' and the other named 'Sizzlean.' They are each supposed to add 6 db at 1K, and 12 db at 10K with compression, but 'Simmering' adds a little less 10K and 'Sizzlean' adds a little more. This was my secret thing on the Maroon 5 record [*Songs About Jane*]: I ran the vocal through both channels. It just really put the voice right up front. This two-channel unit is one of my favorite things ever. I would never use it as a stereo unit, but I would use it as multiple mono, depending on how much top end I would want to put on the signal. 'Simmering' was some high end. 'Sizzlean' was a lot of high end."

Figure 2-36. Maroon 5's *Songs About Jane* album.

Figure 2-37. The mythical Fairchild 670.

Figure 2-38. Eric Valentine's wall of compressors.

the difference by hand. The results are still well worth it. If you are a plug-in person, you've probably used one of the many Fairchild emulators out there. Some of them are pretty good, but touching a real one and turning those big Bakelite knobs—mmmm, that just gets me off!

THE STUDIO STANDARDS

As far as rack-mount compression goes, you can still find the tried-and-true standard units readily available and immediately useful. Try a vintage or modern UA 1176, DBX 160 or

Eric Valentine on Cranking the LA-2A

"The Teletronix LA-2A compressors are incredible distortion boxes, very popular here at Barefoot Recording. Taking a direct guitar signal, running it through an LA-2A and just cranking it up–they are amazing for that. It's my engineer's favorite thing in the world. He runs everything through the LA-2A."

Matt Wallace on the Love of Compression

"Here's how much I love compression: When I'm tracking, we are monitoring through my SSL quad compressor. I'll put a little bit on it just because I love that sound. I'll turn it off to check on things, but I just like that glue thing, and I know it's going to eventually go there, so I start early. For me, compression adds the necessary excitement that you miss when seeing the band visually."

Michael Beinhorn on Compression

"Of the newer tube compressors, one of my favorites is the Tube-Tech SMC2B, which is also relatively fast for a tube compressor, has nice coloration, and is extremely variable. The Fairman and Manley stuff is nice, too. I tend toward transistor/solid-state compressors like the 1176 Rev D because they are generally faster, they can be very aggressive, and still retain lots of detail. In spite of this, I will generally use a compressor not because I like its tonal qualities, but because it lends itself to the sound I'm recording and is applicable to the rest of the signal chain I'm using. Across the two mix, I really like the E and G SSL stereo bus compressor, but sometimes the Tube-Tech SMC2B or a pair of UA 1176 Rev Ds sound good, too."

160X, Teletronix LA-2A, Urei LA-3A or LA-4, or Compex compressor. Manley Labs and Tube-Tech make great new units. Empirical Labs makes the fabulous Distressor, and yes, RNC makes a really nice compressor.

DO YOU EVEN NEED COMPRESSION?

Some engineers insist that compression is not necessary in recording and that it takes the life out of a performance. Steve Albini is one who prefers to not use compression. And who could argue with the fellow who engineered some of the most famous albums on earth, including Nirvana's *In Utero*? I remember reading about Steve's aversion to compression, so when he showed up unannounced to my session with Babes in Toyland in 1997 at Pachyderm Studio in Minnesota, I was embarrassed to have a rack of rented Urei 1176s all plugged in and squashing the drums, bass, vocals, and whatever I could stuff through them. Ouch! After that, I became more conservative with my use of compression for a time, but I admit to still being heavy-handed with laying on the sauce. Sorry, Steve.

SPACES, PLACES, AND TRAINS

Possibly as important as the equipment and the procedure in recording is the recording space. It can be considered as having as much of an effect on the recording as any particular

Al Schmitt on Doing It by Hand

"Back in the day, we didn't have any compressors. So we did a lot of vocal riding on rotary faders. But we rolled gain on them. Most singers had a thing called 'microphone technique,' and they leaned in on low notes, and they backed off on high notes, and they would just lift their head a little on p-pops."

Figure 2-39. Briar Wilson back in the day.

type of microphone or preamp, or perhaps even more of an effect because the space inspires a certain type of performance in the artist. Some of the most interesting recordings have been done in some very unusual places, such as at churches, theaters, castles, temples, caves, boats, trains, and so on. Here are just a few for starters:

CLEARWELL CASTLE

Yes, it is true. Parts of Led Zeppelin's *In Through the Out Door* album were recorded in a castle, and this was it—the Clearwell Castle in Gloucestershire, U.K. As a way to escape the weight of their fame, the band retreated into the woods and recorded in a number of remote locations, where they could concentrate without distraction. Clearwell was a favorite place.

Figure 2-40. Clearwell Castle.

CASTLE RÖHRSDORF

Castle Röhrsdorf is outside of Dresden, Germany, in a rural setting. Once a royal residence, the castle became a school, and then was partially abandoned for many years. Today it

Figure 2-41. Castle Röhrsdorf.

houses a unique studio with a special collection of Neve, Telefunken, and Studer consoles, extremely rare microphones, and delightful spaces.

THE MAJESTIC TAJ MAHAL

Two Paul Horn albums were recorded in this beautiful tomb, built in 1632 near Agra, India, by emperor Shah Jahan as a mausoleum for his wife Mumtaz Mahal. Paul's label and producer/engineer Al Schmitt got permission from the Indian government to record in the evenings inside the magnificent memorial. The space was huge and reverberant, and there were challenges to overcome, including recording around noisy pigeons.

AL SCHMITT'S TAJ MAHAL EXPERIENCE

"I had been Paul Horn's producer at RCA and did the *Jazz Suite* and several other albums with him and Lalo Schifrin—five albums in all. I did one especially interesting record with Paul. We recorded at the Taj Mahal in India. He did two of these records; he had done one earlier, and I went the second time [in 1989].

Figure 2-42. Taj Mahal.

"We flew over and met [Indian Prime Minister] Rajiv Gandhi at two o'clock in the morning to get permission. We talked to him for a while, and he was really an interesting guy. So, he asked what kind of equipment we were going to bring in the Taj Mahal, and I said, 'Well, it's just a little DAT machine.' I had two mics and a DAT. And he knew what a DAT machine was. He said, 'You know, I have one of those!' And I said, 'You're kidding me!' They were new, DATs, so I asked him—'Where did you get it?' He explained that a Japanese contingent came over and they brought it to him as a gift. He didn't know how to use it, he said. So I was going to give him lessons, but he was just too busy. But he gave us permission to record, and we went to Agra, the district where the Taj Mahal is, and we went in and it was just Paul and a cantor. A traditional cantor that sang some nonlyrical stuff.

"Paul had a bunch of flutes: wooden flutes, Chinese flutes, all different. Well, we couldn't go in until late at night to record, because tourists were in there during the day. But then later the army came and took us in, and guards would watch outside to make sure nothing funny was going on. The Taj Mahal is open at the top, and pigeons come in, and we could hear the pigeons cooing. So we had to sit in there and wait for a while—to wait till they died down and went to sleep. And then we recorded. It's like recording in an Echoplex in there, because the decay time was I think at least fifteen seconds. But in the middle of a couple of the recordings, we would hear 'Splat! Pat, pat, pat!' It was bird shit that would come down and 'splat, pat, pat, pat.' It would just echo all over. We would have to stop and wait a while and then try another take. So we thought we were going to be able to get it in one day, but we couldn't, because of the birds, so we went back the second day and finished it up."

Figure 2-43. Willie Nelson's *Teatro* album, produced by Daniel Lanois.

TEATRO

Daniel Lanois had an open-room recording setup at his Teatro Studio in Oxnard, California. Willie Nelson's *Teatro* album was recorded there, with the old deco-fronted theater on its cover. The recording equipment was installed directly in the open theater, right down on the floor in the middle of the room.

RADIOSTAR STUDIOS

I know from experience working at my RadioStar Studios, in Weed, California, that everyone sitting onstage could hear us speaking in low voices in the back of the room. That's because these old vaudeville theaters were made acoustically to have sound from the stage travel to the very back row of seats without amplification. RadioStar's theater had several unique spaces, including an enclosed balcony and the secret "dungeon," which became the favorite recording spot for several producers, including Ross Robinson.

It was in RadioStar's dungeon that Ross recorded From First to Last's *Heroine* album, fronted by the young and powerful Sonny Moore, who later rose to fame as Skrillex. The dungeon's ceiling was only five foot tall, and you had to crouch to get in there. It was right next to the diesel-burning steam boiler furnace, and the heat in that cramped space was oppressive. Cram a bunch of musicians in there with drums, amps, mic stands, and an insane shouting producer—well, you are going to get intense recordings.

AIR STUDIOS

No wonder Sir George Martin chose this impressive Victorian-era church building for the AIR Studio facility, moving the studio there in 1991. More than just a church, this facility (named Lyndhurst Hall) boasts stunning architectural features and one of the largest recording spaces in the world, maintained to the highest quality. Top it off with a beautiful 96-input Neve 88R console and an outrageous collection of vintage

Figure 2-44. Sonny Moore, also known as Skrillex.

Ross Robinson on Norma Jean's Dead Body

"I worked in the dungeon—the catacombs under the stage at RadioStar—because of the intimacy. I made one of the best records down there. Actually, two of them—but the Norma Jean record was extra phenomenal. When we recorded From First to Last at RadioStar, the dungeon was all decked out with Halloween lights. On the Norma Jean record, though—well, they found a dead body. Not in the studio, but up in the snow on Mount Shasta, while they were recording the album. It was apparently a suicide. They called the cops when they found it. Yeah, that place was weird."

Figure 2-45. Norma Jean's *Redeemer* album.

Hans Zimmer on Using Churches to Record *Interstellar*

"The recording for the film *Interstellar* was done in this amazing church in the middle of London. It's the Temple Church, the church built for the Knights Templar as their headquarters in the late 1100s. It was the church that appears in *The Da Vinci Code*, and director Ron Howard shot that film in it. In *The Da Vinci Code*, that particular church was known as the 'center of power,' and when you are there, you realize it really is! As a lawyer, there is a saying: 'You have been called to the bar.' Well, actually, 'the bar' refers to the altar at the Temple Church. We set up a mobile recording system and recorded part of *Interstellar* in there because it has an amazing pipe organ. The other half of the score was recorded at AIR Studios, which is also in a Gothic church. AIR also has a pipe organ, but it doesn't work. It's actually filled with sand, because its pipes are a tuned resonator, if you think about it, so it is going to fuck up everything if you record in the same room with it just sitting there!"

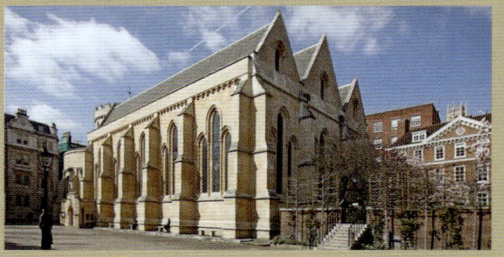

Figure 2-46. Temple Church, London.

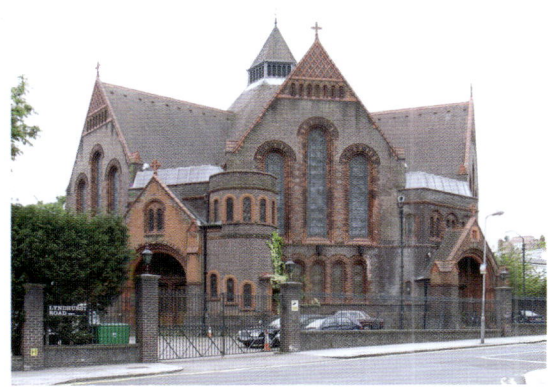

Figure 2-47. AIR Studios, Hampstead, London, U.K.

recording finery. Sir George no longer owns AIR Studios—he sold it in 2006—but what he built continues to attract the top echelon of the music industry, including Coldplay, the Black Eyed Peas, and Mumford & Sons. Located in London's Hampstead district, it is one of the world's finest studios, all build around an old church!

HEADLEY GRANGE

Headley Grange was an unlikely spot for Led Zeppelin to record, but they needed a remote place to concentrate on their *Led Zeppelin III* album. So good were the results from their time at Headley Grange that they came back to record *Led Zeppelin IV, Houses of the Holy,* and *Physical Graffiti.* Genesis wrote *The Lamb Lies Down on Broadway* there, and albums by Bad Company and Fleetwood Mac also were spawned in this place. Located in Hampshire U.K., Headley Grange was really just a broken-down old rooming house, but was a perfect getaway where the artists had few distractions. That is where its success mattered. No equipment was built into Headley Grange—most recording was done with the Rolling Stones' remote truck. This was where Led Zeppelin's famous "When the Levee Breaks" drumbeat was recorded, when Andy Johns put a pair of mics at the top of a stairs while John Bonham played a kit set up at the bottom of the staircase. For you gear nerds, that famous drum track used two Beyerdynamic M 160 mics.

Figure 2-48. Jimmy Page.

Shelly Yakus on Alice Cooper's "Billion Dollar" Mansion

"We recorded the Alice Cooper *Billion Dollar Babies* album in a mansion in Connecticut on a dead-end street. The place was on the Galesi estate, right down the street from where Bette Davis lived. It had a fireplace in the ballroom so big you could drive an SUV into it. Beautiful frescoes in the ceiling. Bob Ezrin produced, and I engineered. We set the Record Plant remote truck up in the driveway of the mansion and ran the cables inside. We made a great record there, but it was like being in a hotel, because between every take, the band would disappear to their rooms on the second floor. It would take an hour to get them all back down again."

Figure 2-49. Alice Cooper's *Billion Dollar Babies* album.

Bob Clearmountain on Recording Residential with Bryan Adams

"We recorded Bryan Adam's *Into the Fire* album at his house in West Vancouver. He had just moved in, so he had no furniture in there yet. He had just put an SSL 4000 in the basement, so we put the drums in the dining room, the guitar amps and a Leslie cabinet in the bedrooms, and Bryan, Keith (lead guitar), Dave (bass), and Tommy (organ) in the living room, along with the drum room mics, as it was open to the dining room. There were no mic tie lines, so we ran the mic cables down the ash chute in the fireplace to the SSL, and put up video cameras, so I could see them and they could see me. I loved the way the recordings sounded."

Figure 2-50. Bryan Adams's *Into the Fire* album.

Figure 2-51. Peter Gabriel's Real World Studios.

REAL WORLD STUDIOS

A beautiful studio compound built into an old mill in Wiltshire, U.K., with the main building surrounded by a serene millpond. Outrageous equipment is spread throughout several

Bob Ezrin on Phish's *Farmhouse* Album

"Phish's barn is a perfect acoustical environment: natural, not too reverberant, and rich, as it's totally made of old, porous wood. So recording there was easy and sounded great. The only challenge in the barn is isolation, but with these guys, that's not an issue. In fact, I loved the sound of their instruments bleeding together in the way they would in any big room. For that reason, we chose to record in an older, wooden room in Nashville (Black River Entertainment) where Orbison used to record when they were there. I had them somewhat separated, but not overly, as most of the performance on the album, at least for the main tracks, is them playing together. We did very little overdubbing, comparatively speaking, outside of vocals."

Figure 2-52. Phish's *Farmhouse* album.

ENGINEERS AND DEMOLITION EXPERTS

> ### Ron St. Germain on Killing Joke in an Egyptian Pyramid
>
> "While mixing the Killing Joke song 'Pandemonium,' I heard a reverb effect in the recording that I had to stop and ask them: 'What the hell is that?' I literally pushed back from the console when I heard this crazy howling sound, like banshees on fire! Bassist and producer Youth said it was the 'power of the triangle.' In their quest for originality, Killing Joke had rented a pyramid in Egypt and had gone up into the apex with a DAT recorder, got naked, lit candles, and proceeded to run around screaming and singing random lyrics, inspired by the energy of the moment."

studios in a facility that also includes a community kitchen, residences, rehearsal and production space, and a record label. One of the most beautiful spots on earth to make music, for certain. The brainchild of Peter Gabriel—someone with extremely good taste!

WAYNE COYNE'S PARKING LOT EXPERIMENT

Flaming Lips frontman Wayne Coyne gathered thirty folk from his hometown of Oklahoma City to participate in a shared musical experience, dubbed the "Parking Lot Experiment," in 1997. What developed out of this became *Zaireeka*, a four-CD recording designed to be played from four separate stereo systems simultaneously. But the initial event needed some tweaking to get it to release-ready condition.

In the original gathering, everyone drove their cars into a local mall parking garage. A prerequisite for participation was that you had to own a car with a cassette deck. Wayne provided each participant with a separate cassette tape and had everyone start the tape at the same time. Wayne directed the start of the experiment with a bullhorn as he ran around the lot. After a few attempts, the tapes were aligned and playing, and what people heard was basically sixty tracks all playing simultaneously, all coming out of separate speakers in a large reverberant cement room! It was fantastic!

Participants strolled through the parking lot between cars and marveled at the experience. So astounding were the results from the experiment that Wayne next took it to the South by Southwest conference in Austin, Texas, in 1997 and similarly performed the experiment for a crowd of one thousand people.

In New York, touring the experiment was a challenge because there were no parking lot venues available, so Wayne solicited for thirty boom boxes to perform the parking lot piece—also with spectacular results—in a bar. So how could this be translated into a commercial release? Here's where the next level of genius comes in. As a Flaming Lips release, the "Parking Lot Experiment" was submixed into eight channels with mixer Dave Fridmann, then packaged into four stereo CDs to be played in four CD players simultaneously. You can do it yourself with four car CD players, or four boom boxes, or any combination thereof. How about four iPhones?

"STATION TO STATION"

One of the most incredible studio tales I've ever heard was on Doug Aitken's "Station to Station," with a studio on an actual moving train traveling coast to coast across America in 2013. Developed as a massive art project, the train was scheduled to make nine scheduled stops on its four-week journey, and at each stop there was to be a festival "happening," with art, film, and music. Producer Justin Stanley—who had done quite well with the Vines, Sheryl Crow, and Eric Clapton—made the suggestion to the art director at Levi's, the "Station to Station" project's sponsor, to build a recording studio into the art train, and everyone agreed!

Figure 2-53. The "Station to Station" studio in a rolling train.

JUSTIN STANLEY'S ON-THE-RAILS ODYSSEY

"The train was about seven cars. It was beautiful. There was a storage car, two accommodation cars, Levi's had a car full of denim. There was a media car that was full of computers, and for the trip they had at least five Alexas—the film cameras, more cameras even than a motion picture—and they had ten Canon 5Ds, so they were filming everything. And they had edit bays in the media car and were editing film as they were going. And then in the next car was the recording studio.

"When I got to Washington, D.C. in advance of the trip, I arrived to a blank car, a blank canvas. The whole thing was totally empty. And I had three days to build something, to put a studio together. So first I put down a wooden floor because it was a classic old train car from the '50s and kind of rickety. Then I created a whole system where I Velcroed or bolted everything to the floor—all the amps; the drum kit was bolted to the floor. All the keyboards were strapped down. Everything was attached basically so it wouldn't move around. That was a major bit of work. And then as far as gear goes, I brought a laptop and an extra screen, a Symphony converter by Apogee and a UA Apollo. There was a little Rupert Neve centerpiece, a bunch of mic pres that I had at home, compressors and whatever, and a whole bunch of different kinds of mics—simple mics, nothing too crazy.

"I really wasn't concerned with outside noise problems, because I really thought it was going to be noisy as hell. In fact, I tried not to have any preconceived ideas on what was going to happen or who was even coming on board. So we started out, and the first person to show up was Thurston Moore from Sonic Youth. It was Thurston and his drummer—they sat down, and we just started recording a song. We were on an old piece of the railroad track, and we were listening to the clackity-clack of the train. So he wrote a song around that,

Ian Rickard on Recording in a Cave

"Pluto's Cave is pretty much in the middle of nowhere. With my dog leading the way, we all trekked down the trail and climbed into the huge dusty volcanic cave looking for 'the spot.' We used a laptop and a USB converter with two inputs—one for the vocal mic and one for a mic to capture the natural reverb of the cave. The way the rays of the setting sun were shining through the ceiling of the cave was majestic and very inspiring. It didn't really matter that none of the takes actually made it as lead tracks—we used many of them as background vocals. And what a memory it made!"

Figure 2-54. Ian Rickard recording vocals in Pluto's Cave, California.

which was the perfect start of the whole trip.

"Outside noise was not an issue. I just took it all in . . . I actually wanted more! So we would hang a microphone outside sometimes. Really weird, right?

"There was at least one stop a day. Major cities like Chicago, Illinois, and Pittsburgh, Pennsylvania, and a few little stops in towns like Barstow, California, and Winslow, Arizona. So at almost every stop, we had a festival concert that I would record and mix it that night. When we stopped at Winslow, Jackson Browne played, because he cowrote that song 'Take It Easy' by the Eagles, which uses 'Winslow, Arizona' in the first line of the lyric. Beck did a concert in Barstow, California. They had a flying saucer flying by, hung from a helicopter—it was surreal.

"Mavis Staples was at another stop. During the day while we were traveling, we'd record whichever artist was on the train. Ariel Pink stayed for two weeks on the train; Cat Power also rode along for a while. I didn't get a lot of sleep—it was the best time ever. Not only were you feeling the train's movement with this massive engine that was carrying you to the next destination, but you had this incredible view out the window that kept changing. So you were constantly being inspired by your surroundings. It was a really 'moving' experience.

"The greatest thing of all about the whole 'Station to Station' adventure is that it really brought the art of simplicity to the whole process. You don't need a lot to make the right record. You need a vibe. You need people who are into it."

Recording in a Silo

Leave it to some crazy community-minded artists to make a grain silo into a completely outfitted recording studio and public performance space for the residents of Sheboygan, Wisconsin. Built from parts of a typical Wisconsin silo and salvaged bits of a camper truck, it's called M.I.K.E.–the Musically Integrated Kiosk Environment. It's totally unique and loud as hell in a rainstorm!

Figure 2-55. Meet M.I.K.E. (Musically Integrated Kiosk Environment)

BOB EZRIN ON DAVID GILMOUR'S ASTORIA

"I love that studio. It's a Victorian concrete barge that was built for Fred Karno, the great West End impresario, at the turn of the 20th century as his private 'knobbing shack,' and moored near Hampton Court, down the Thames from London. It was designed as a place where he could bring his latest ingenue in privacy or host soirees with the beautiful people of the time.

"Charlie Chaplin was a frequent guest on the boat. It is opulent and spectacular in that Victorian style, with lots of ornate fixings. The top deck has a bandstand and dance floor with magnificent lighting all round. At night, the boat looks like something from a Victorian amusement park . . . and yet it stands behind a brick wall that shields it from the road, so no one can see it unless they enter the grounds. It sits at the bottom of a luscious, rolling lawn, at the top of which is a conservatory, which has a full kitchen, eating and lounging areas.

"The studio itself has been lovingly inserted into the boat without disturbing its structure or any of the original fittings. All the original rooms are intact. And all the technology sits on the original floors, which were gently lifted to create conduits for cabling, etc., and then replaced. The 'services' and HVAC, etc., sit outside on the lawn, but hidden inside a trellis that extends out to the boat like an arboreal bridge. It looks like natural, riverside growth and doesn't disturb the tableau at all. It's a remarkable piece of engineering and repurposing, really. I think David Gilmour and his team, headed by Phil Taylor, did an exceptional job in preserving the integrity of this magnificent artifact while making it a fully functional, best-of-breed studio.

Figure 2-56. David Gilmour's Astoria.

"As well, I find that the river itself infuses the work we do there. It's sometimes joyful, like when the kids row their sculls past us as we work; sometimes brooding, like in the fall and winter. And sometimes it has a primal energy, like rolling thunder. But it is *never* neutral. So every day working there is a new experience. I wish I could work there more often, but it's a bit of a drive from Nashville."

BOB CLEARMOUNTAIN ON RECORDING ON TOP OF A MOUNTAIN

"I recorded a live video ad for the State of California featuring the band Band of Horses on the top of the mountain, just above the 'Hollywood' sign. They had two vocals, two acoustic guitars, mandolin, upright bass, drums, and a string quartet. It was quite tricky, as there was very little room up there to set up, along with lots of tourists hiking up there, stepping all over everything to get a look at the view.

"We recorded on a MacBook Pro laptop in Logic and had a couple of Apogee Symphony I/O interfaces with preamp cards. The city would not let us get power from the fenced-in

Figure 2-57. Band of Horses.

microwave antenna array right next to where we set the recording gear up, so the producer rented battery packs for us. The real problems started when the battery pack the converters and hard drive were plugged into went dead in the middle of a take! We scrambled to plug in another one, but as anyone who has had the power go out on a hard drive while recording knows, getting it going again was a nightmare!

"This was a problem, as the sun was setting and they were trying to get just the right lighting—plus, they had booked a helicopter flyby with a camera at a specific time. We actually ended up getting one good take, which I had to do quite a bit of processing on to remove the wind noise. Many people who heard it were sure it was prerecorded in a studio and they lip-synched, but it was entirely recorded live up there."

Ross Robinson on Sepultura's Canyon Percussion

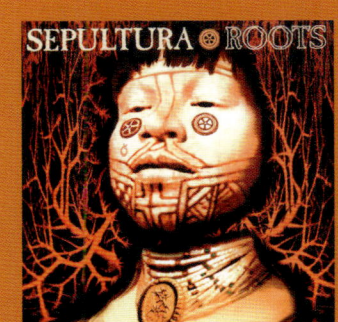

Figure 2-58. Sepultura's *Roots* album.

"We got three football fields of cable and set microphones all the way down the bottom of a canyon and up the other side at Indigo Ranch. It was at that big rock that I had the band Sepultura play percussion on Roots. The echo was dramatic—'Bam! Ba-bam!' shooting around in triplets. It was kind of a mess, but it was cool as shit. The sound at the beginning of that album is those mics set up on the cliff there in Malibu. You hear all the insects, and you can really catch the vibe of the place as the sun went down at night."

Figure 3-1.

3

ALL THE WORLD'S A STUDIO

Figure 3-2. Engineer Ian Rickard, just another day in the office.

OPENING A SCENE

Now that you've gotten a glimpse of my viewpoint on recording, let's start looking at some ideas that will help unhinge your own perspective. Something to make you rethink your own recording strategies! First of all, take a moment to gather your senses . . .

THE BRAVERY OF LIVE-TO-TWO-TRACK

One of the boldest types of recording is live-to-two-track.

Larry Crane on Hot Dog Delirium

"I recorded a guy in a hot dog suit once. He was in a band called Frank Furter and the Hotdogs. But it was really just one guy. He played guitar and a kick drum, and his songs were like 'Baby, Don't Treat Me Like a Corn Dog.' I worked with this guy at the end of a long run of thirty days, sessions going ten hours a day. I'm just exhausted. My hands are starting to twitch. We start the session and he puts on the hot dog suit, and I had to rub my eyes. 'Am I really seeing this?' I thought I was losing my mind."

This is where mics are set up, processed, and mixed at the time of performance. This takes plenty of commitment by the people involved to "get it right" at the time of recording. There is no going back—the mics are all summed to two-tracks. The trick is to record as many takes as needed, then cut together the best parts from each take during editing. You can even work on each section of the song separately, adjusting the levels and effects for that part of the song. If there are panning and live mix moves to do while recording, you can rehearse each section, one at a time. Take several passes of each section until you achieve the desired take, and then assemble with editing. Frank Black, the genius behind the Pixies, is the master at this type of recording, having done at least four albums with this approach. It is immediate and captures the interaction between the players. A tricky thing to do. As a band, you really have to be ready for this technique. There is no going back and fixing it in the mix!

Figure 3-3. The Magnificent Ed Stasium.

Ed Stasium on Mick Jagger's All-Star Band

"When we went to New York to start mixing his solo album at Right Track Studio A, Mick Jagger decided to use the studio space for rehearsals for an upcoming tour. He assembled an incredible all-star band—Jeff Beck on guitar, Simon Phillips on drums, Doug Wimbish on bass—and they'd break into these incredible jams. I recorded it all to digital two-track, and because no one was wearing headphones, the recording did not get in the way of the performances at all."

Figure 3-4. "The Hammer" and EM-EQ2 stereo EQs by A-Designs.

ULTIMATE EQ COMMITMENT

Live recording requires extreme confidence, with the results being instant and final. But it doesn't have to sound like a rehearsal recording. Using a stereo EQ will give it a finished "mastered" feeling that you appreciate immediately. Don't be a pussy with your stereo EQ. Push the mids to make it honk for a vintage sound. Or scoop out the center for a polished hi-fi vibe. Push the bass to make it boomin'. Add air on the top to give it

Ross Hogarth on Wall-to-Wall Porn

"I produced this heavy metal band called Coal Chamber, and Dez, who is the lead singer, is way into all this satanic stuff. So I let him build a satanic chapel for himself in the iso booth. So when he went in there, he felt like he was in his own world. But on the flip side, the guitar player, Meegs, is way into porn. I don't know if he still is—he's married now—but back then, he was super into porn. He set himself up in his iso booth with his guitar rig and left for a little bit. So I sent the runner out to the corner porn store to literally buy fifty magazines and cut all this porn out, and we wallpapered the whole iso booth with it. Like wall-to-wall porn. And when Meegs came in, he loved it."

Figure 3-5. Coal Chamber's porn room.

Jim Messina on What Is Really Important

"As a writer, every word has its place inside the emotion and phrasing of the lyric. And the same goes for the song's recording. The sound of the recording is important, for sure—but I would gladly trade the ownership of a technically perfect recording of a bad song for a one-miked, mono recording of a great song recorded live, in one take on an old mono 300 Ampex."

Figure 3-6. The Ampex 300 mono tape recorder has one big meter.

Al Schmitt on Live-to-One-Track Ray Charles

"You know when there are certain records you're working on that are just going to be classics. The one I did with Ray Charles and Betty Carter was one of my all-time favorite records. And we did the whole album in like seven and a half hours! All done mono and no editing, nothing! He played piano and sang; she stood right next to him. I had a U 47 on her and a U 47 on him, had mics on the piano, and then there was a full choir and a big band! Some of the sessions even had strings! And everything was recorded at the same time in mono—and that's how we learned to record! Everything was done at once!"

breath. You can really give your two-track recording a "shine" to make it instantly exciting! The manufacturer A-Designs makes a stereo EQ rack unit called "The Hammer" that really digs in for effect.

ANALOG WHEN YOU DON'T HAVE TO

Yes, analog tape is a pain in the ass. It's been decades since the acceptance of digital recording in the recording industry, and I've pretty much moved on from the cumbersome idea of loading up and punching in on rolls of tape. However, there is a sweet effect from tape compression that is not reproduced by any other method. While working with Prince on mixing his *Diamonds and Pearls* album, I noticed that even after a proper machine alignment, the needles were pegged on every meter on the Studer A800. "Pegged" is the term describing the indicator needle bouncing off a little steel peg on the face of the VU meter, as it tries to

Eric Valentine on Using Tape As an Effects Processor

"Most of the time, I am capturing the recordings both digitally *and* through analog tape machines simultaneously. I have a variety of old tape machines: an old tube Ampex tape machine, an MM-1200, a Studer J37, a Scully 16-track. I use the tape machines more as 'effects processors' than as a way to store the performances.

"Basically, to start a session, I will record a 'test' performance digitally, then give the band a break while I set up the analog recorders, playing back the digital test take into the tape machines to adjust the routing and analog input settings. Then, when that is finished, we will do the actual recording, of which I record two sets of tracks in Pro Tools, both the direct digital tracks that the band hears while they are playing and the analog tape tracks that are not monitored until playback. Depending on what the aesthetic is for the project, I may hit the tape hard or use the tape in other ways."

Figure 3-7. Eric Valentine's machine room.

Brad Wood on Analog Tape

"If you are a smart engineer and you choose to use analog tape because you know what it's doing and you like what it's doing, fine. There is no right or wrong about it. Some of the best records made were done on the most screwed-up limited analog gear ever. I think that limits and structure are more often than not an ingredient missing from digital recording."

handle maximum level. It was a cause for concern, but after getting into the mix, I realized that there was a certain sound that over-recording caused on the master tapes, and it worked for Prince's music. In fact, it gave the tracks a certain "excitement."

OTHER ANALOG ALTERNATES

Another nice side effect from driving tape machine electronics harder than prescribed is a type of mild saturation that is being emulated in the digital world. A company named Burl makes digital converters that use the same transformers used in Ampex tape machines, nicely rounding off some of the harsh digital edges that make me so uncomfortable.

So how do you get analog tape saturation in real time, without actually recording your tracks to tape and transferring to digital? There is a device that actually does it called C.L.A.S.P., created by the folks at Endless Analog. But really, folks, at this point does it really

Eric Valentine on Creating a "Vintage" Tape Sound

"There are sections on the project I'm working on now where we are going for a really sort of vintage '50s Del Shannon kind of sound, so I'll run the recording through multiple generations of tape. On the way into the computer, I'll split the mic signals, literally sending it into one track of the tape machine, out of that track into another track of another machine, and then maybe in through a third machine, maybe the Studer J37, because what I am trying to achieve is the end result of that Del Shannon recording process. They would have to capture a performance on a particular track, then submix that to another tape machine. You are potentially going through two or three tape generations in order to get the sound we ultimately hear as the finished song. Increasing the noise floor has never been a problem for me, because a lot of times I'm running the signal quite hot on the way in to the machines, something they had to do as well on those old recordings."

Figure 3-8. Eric Valentine's '50s-style tape device.

Damon Fox on Staying Inspired

"When you're in the moment, well, it's about quickness. Whatever keeps you inspired and gets the performance out—that's all that matters to me. Just the performance. If it happens on analog gear, fantastic. If it happens on GarageBand, I don't think anybody gives a shit. I don't think it matters anymore if it is sampled or emulated. If the hook is good and the arrangement is good, but it's been recorded to ADAT—well, you've arrived at the same place. It's too late to be a crusader."

Susan Rogers on Vari-Speeding Prince

"Vari-speed is an adjustment setting on an analog tape machines that is very useful to get a timbre of an instrument that you want, that is not its native timbre. Prince used vari-speed a lot in his music. He would vari-speed a tape recorder down a lot to get a higher timbre or the voice, and particularly on his guitar."

matter? I think the days of tape are truly behind us, but if we can have fun cranking up the old tape machine, let's do it. Spending a ton of cash to get that sound, though? Not sure it is really worth it. I'd rather spend that money on hiring a private jet to record my next album on, with laptops and Mboxes twenty thousand feet in the air. It wouldn't really sound much different than if it had been recorded on the ground, but what a great story it would make!

RECORD WITH WHATEVER!!!

Go ahead and record with whatever you have! With an iPhone or iPad, take video and use

Hans Zimmer on Recording with the iPhone

"When we were doing *The Dark Knight Rises*, I read the script and said to Christopher Nolan, 'You know, I think this scene needs a hundred thousand people chanting!' He said, 'That's a great idea!' So at one point in the production of the movie, using social media, we invited thousands of people to be a part of the movie! With the help of Warner Bros.' brilliant media genius, we reached out to have a hundred thousand people basically record themselves on their iPhones or on their computers or whatever, and send it in to us. I realized that on every chant, the acoustic space would be different! I thought, *Oh, this will be really interesting*. But it wasn't at all—it just became this huge, big mess! Oh, yeah, there were many engineers here for days and nights, cutting and editing takes, and in the final version there were literally thousands of people chanting. And weirdly enough, it sounded too good!"

Figure 3-9. Hans Zimmer.

Ron St. Germain on Smoking Bibles in Prison

"Singer H. R. had to bail out of the Bad Brains sessions on *I Against I* because he needed to go to prison for a pot possession conviction. Big surprise right in the middle of a record project. Before he had to leave, we managed to get all the vocals on the album completed, except for the song 'Sacred Love.' So now what was I supposed to do? H. R. was in Washington, D.C.'s Lorton County prison and we needed to finish this record! Eventually, after more than a dozen calls, we were able to arrange permission from the prison officials for H. R. to phone in his vocal performance. We coordinated a time to do it, and H. R. called in collect to the studio. I accepted the charges. The studio had a direct-patch phone system, but of course it decided not to work. (Insert expletive.) I scrambled to save the session by duct taping a Neumann U 87 to the earpiece of the phone, and wrapped that with a moving blanket to isolate it from control room noise. I used a Teletronix LA-2A to gain up the tiny telephone signal to get a decent level to tape. In order for H. R. to hear the track, we sent the mix through an Auratone speaker and basically gaffered that to another telephone handset mouthpiece. So we were set, but H. R. was not quite ready to perform. He needed to get in the right 'head space,' so he reached for his Bible. He always carried his personal Bible everywhere, and was allowed to bring it into the prison. Turns out, he would use a specific page from this Bible as a tray to clean his herb, and after years it was laden with resin. H. R. had anticipated that someday he would need it, and this was that occasion. He actually tore the page from the Bible, rolled it up, and smoked it before we recorded what was to be one of the most infamous performances of H. R.'s career, and one of the most memorable studio experiences in mine."

Figure 3-10. Chris Johnson phoning in a solo.

the audio from it. Lots of kid's toys. Old cassette decks from the thrift store. Never miss an opportunity to try something cheap and new, or old and weird!

LOUD NOISES AND SMASHING THINGS

The island volcano of Krakatoa exploded off the coast of Indonesia in 1883, and the sound was heard around the world. Seriously, the sound wave was reported to have circled the globe seven times. Well, glad I wasn't around for that. OK, then:

What is the loudest man-made (or woman-made) sound ever recorded? Hard to say, but I wanted to make a very loud noise and have fun doing it. My first thought was to take a piano and drop it from a tall building. Oh, what a cacophony that would be!

Well, after failing to find a building owner to be a willing participant, I investigated dropping a piano from a crane. Pianos are easy to find—after they reach a certain point of disrepair, they are considered totaled and not worth fixing. Dozens of derelict pianos are just given away every day, so I had no trouble finding subjects for my experiment. Crane companies, on the other hand, were expensive and uncooperative, not enjoying my sense of humor or adventure.

We scheduled Tool into Grandmaster Recorders in Hollywood off of Cahuenga Boulevard for their *Undertow* album. Grandmaster had a large, unfinished garage in the back of the studio. It had tall ceilings and was a great place to make noise outside of the formal recording rooms of the studio. I acquired pianos for the experiment, purchased two sledgehammers, a pick, an ax, gloves, and safety goggles. I started with one piano, miking it with acoustic guitar glue-on pickups, then close-miked the piano with a Shure SM57. These $100 mics are known for their ruggedness, so it was placed in the line of fire. Lastly, I set up a pair of Neumann U 87s about ten feet away from the piano. During the recording, drummer Danny Carey also carried in his sawed-off shotgun.

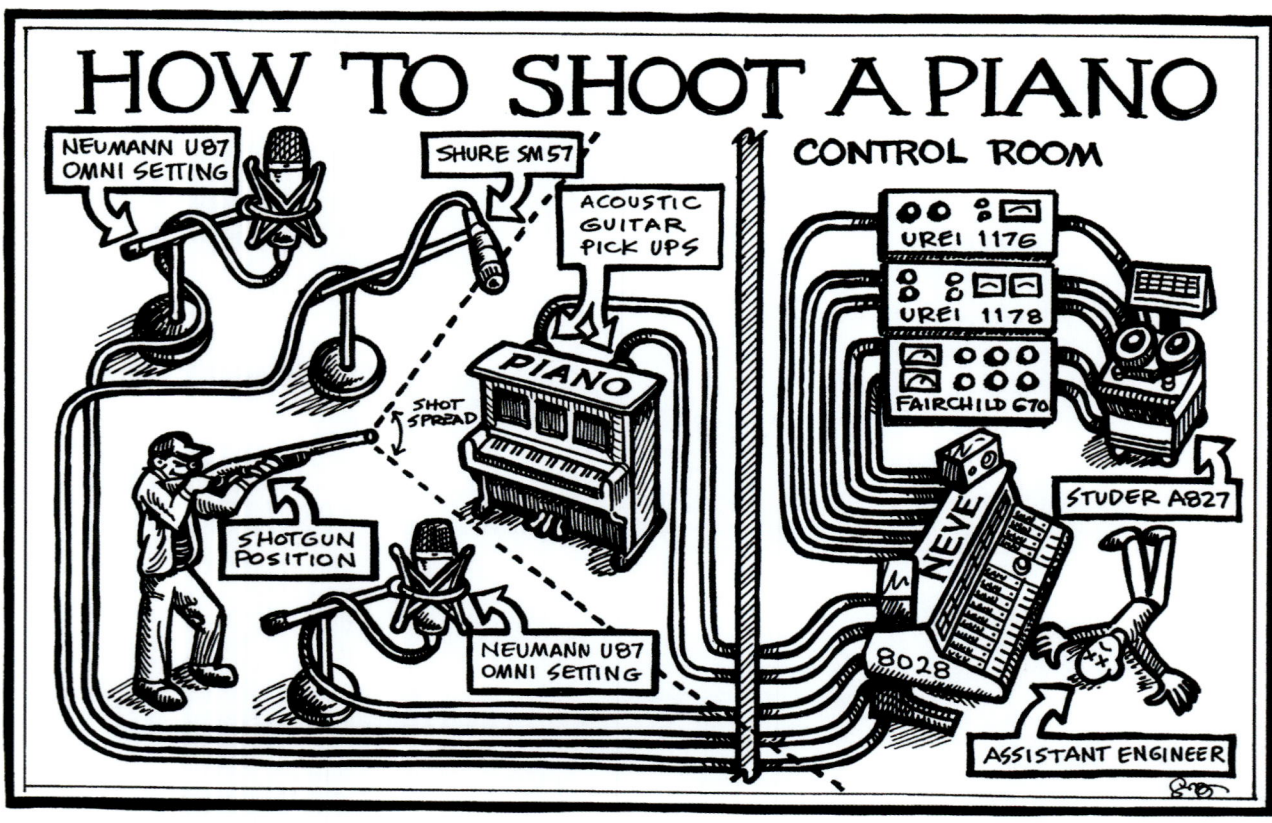

Figure 3-11. How to shoot a piano.

Figure 3-12. Piano smashing with Tool.

Everyone wore headphones, which also served as hearing protection. We set out to destroy, first blasting the piano with a shotgun, which caused the piano to vibrate intensely. Then the band members took turns smashing the piano to bits with the sledgehammers. Recording it all on a two-inch Studer in the adjacent room. The poor engineer in the control room nearly got hit when someone with lousy aim shot out the 57 and also put a hole through the back wall. The recordings from the piano smashing were spectacular. We created a whole sequenced piece using sounds from the piano recording, the band calling it "Disgustipated."

Ross Hogarth on David Lindley's Broken Solo

"When I was David Lindley's guitar tech, there was a slide guitar solo on the *Hold Out* record, where the great distorted sound is literally the tubes in David's amp burning up. The whole amp is just going whack—the way it would sound with a Variac when you dial back the voltage and everything is at some crazy bias. And that solo had to be kept because, well, Lindley could play it perfect every time, but that tone on that one take was just so amazing. I remember Jackson Browne and, God rest his soul, Greg Ladanyi asking 'What's going on?' and I had to go take a look up underneath the amp. 'Agh, tubes are all gone, just fried.' So we put new tubes in, and of course it was never the same, so we kept the broken take."

Susan Rogers on Tape Chewing

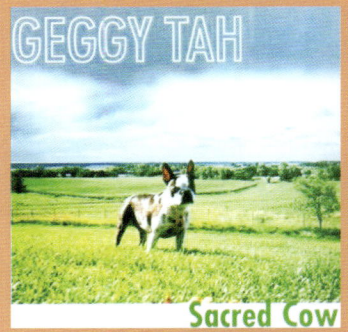

Figure 3-13. Geggy Tah's *Sacred Cow* album cover, featuring Susan Roger's dog Gina.

"On the song called 'She Withers' by Geggy Tah, Tommy Jordan—the lead singer, writer, and leader of the band—wanted the sound of something 'withering' right before it goes into the bridge of the song. So we took the mix and printed it onto cassette. Then Tommy pulled out a piece of cassette tape right in the section of the song we wanted to 'wither'—about a foot long—and crumpled it up, put it in his mouth and chewed it, took my little dog Gina—a little Boston terrier who is on the album cover—and tried to put the tape into the dog's mouth! Just literally withered the cassette tape! Wound it back up, put the cassette into the machine, and played it back. Took the output of that and printed it onto a track. *Scooo*, we would have this whole hi-fidelity song just suddenly literally disintegrate right before the bridge, then pop back to normal full fidelity! It was a great idea! And it worked beautifully, except the record label wouldn't let him do it. So it never wound up on the record."

STUDIO PALS

Figure 3-14.

Martha at Sound City
(Phil Brewster)

Satchmo
(James Saez)

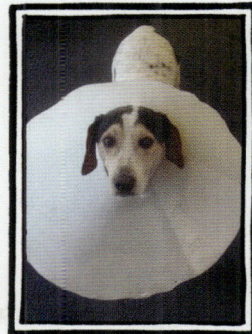
Woody
(Sylvia Massy)

Meow
(Mark V. Meyers)

Woof
(Garth Richardson)

B.C. at Criterion
(Mark Williams)

Memphis
(Perry Margouleff)

Felix & Roxy
(M. Stukenholtz, R. Jacobs)

Bella
(John Norten)

Arthur & Stuart
(Peter Jay)

Boo & Gordy with Bruce Swedien
(Mark Emery)

Figure 3-15. A gallery of studio dogs.

Figure 4-1.

4

VOCALS

Figure 4-2. Singers need attitude, and Rayne Bangsburg brings it!

SVENGALI VOCALIST MIND CONTROL

One of the greatest studio challenges is inspiring the singer into delivering an amazing performance. I do this in many ways. First, I like to have direct communication with the singer, often sitting in the same room with headphones while recording, so there are no secrets. I will sing a part as reference to the singer, but I am careful not to sing it well—I want an insecure singer to think, *Well, I can do better than that!* Next, if I hear straining in the singer's voice, I will distract them away from concentrating on their throat. I'll ask them to use their hands, for instance, to punctuate the lines—the way Italians might use their hands to express themselves.

   I will also ask the singer to really give me the meaning of the words. This is especially good if the singer knows the words by heart. You might ask a singer to memorize a song's lyrics the night before they sing, or even weeks before the session. This will insure a less stilted performance that doesn't sound like it was read off of a lyrics sheet. If your singer can't help but read the words, try taking away the lyric sheet entirely, so that all they have is their performance—no crutches to lean on.

Linda Perry on Heroic Vocals

Figure 4-3. Linda Perry and her Neve.

"Singers get in front of the mic, and they're immediately thinking they should come out like a hero—crushing everything and everybody around, because they are singing full-on and not compressing themselves. Like with guitars, smaller amps sound better than the big ones. You get this centered, fat sound from this tiny little amp because there is no blowout, right? Same with singers. I tell them to get in front of the mic, use their stomach—stop trying to go for volume, but go for power! Power is the richness and the air behind the note. Not the volume. That changes the whole perspective of getting the girls out of that high, nasally noise that happens into more of this richer, deeper-throated, sexier, darker voice."

Matt Wallace on Connecting to the Moment

"When I start a vocal session, I'll have him or her run down the song and sing it once or twice, and I'll get my levels. Then, I ask, 'Hey, what made you write this song?' Then the singer will start to tell the story: 'Well, I was going out with this girl—*bah, bah, bah*.' I respond, 'Oh, man, that's too bad,' and I hit record, and I make them sing it right then. I can actually hear it in their voice when they tell the story: 'Yeah, I was really bummed out when this thing happened.' And at that moment, that energy is sitting in their body and you can capture that feeling. No need for me to even say anything. It's a palpable difference. People are connected to the moment they wrote the song."

WARM UP OR YOU MIGHT FALL DOWN

If a singer has trouble hitting the high notes, try having them move up to even higher notes, even if they lose vocal control and break up a bit. Of course, if it hurts the singer, don't let them continue with this exercise. But what I've learned is that if you show them they can reach even higher notes, then the note you want becomes much easier to reach.

Warm-ups are an excellent way to start a vocal session day, but caution: If a singer does too much warming up, they will not give you that raw emotion that you may want. Sometimes you want a performance that does not have control—a performance that has vulnerability, emotion, pain, and fear. Those first few takes of the day more often hold this type of feeling.

In the same way, depending on the style of music, you may want a vocal performance with anger or bewilderment or deep sadness. I suggest you prepare your singer for that type of performance by forcing them to do something they don't like, or reminding them of a sad story. Especially if the song lyrics contain a painful story from their own life, make sure they are immersed in that story, really delivering the words as if they were talking to you directly.

Figure 4-4. Ice cream truck vocals with Tom Anker.

I'M PICKING UP GOOD DISTRACTIONS

Only one person typically gets in the way of a great vocal performance: the singer. Many singers are self-conscious about their voice not working properly, concentrating too hard on the feeling in their throat. And who can blame them? It's not like they can change the strings and retune, or try a different instrument—they are attached to their instrument. Poor singers are just so worried all the time; I find the best treatment is to get them out of their head. Here is how I do it: I give them so many other jobs to do during the vocal recording that they are completely distracted from thinking about their throat! Some of the distractions I throw at them:

- Stand on one foot while you are singing.
- Pretend you are driving a car while you are singing.
- Sing with your eyes closed.
- Hold your hands together like you are praying.
- Think of your favorite recipe while you are singing.
- Punch the air while hitting the high notes.
- Imagine you are encased in a yellow light as you sing.
- Keep your chin down and your shoulders relaxed while singing.
- Wear boots three sizes too big while singing.
- Wear sunglasses while singing.
- Sway back and forth like Axl Rose as you sing.
- Pretend you are hugging yourself as you sing.
- Keep the air coming out of your lips at an even amount while singing.
- Sing quietly, but with importance and urgency, as if you are right next to a person's ear

and can't sing loudly or you will hurt them.
- Make up brand-new lyrics as you go; completely ad-lib new words on the spot.
- Break away from the melody and create a new melody right on the spot. Don't think about what it will be—just do it.
- Touch your face and the sides of your head while singing.
- Hold an object and sing to it like it is a person.

All of these random instructions may not directly help a singer with their technique, but what they do is keep the singer so busy that they completely forget that they can't sing!

DISTRACTIONS

SING UPSIDE-DOWN

TURN OFF THE LIGHTS

KNIFE JUGGLING

TAPE SINGER TO THE WALL

SING IN A DINOSAUR SUIT

ALIEN ABDUCTION

TURN ON THE CAMERAS

BRING IN A LADDER

SHOCK THERAPY

SING NAKED

EXCESSIVE EXCESS

SUBMERGE VOCALIST

Figure 4-5. How to torture a singer.

Shelly Yakus on the Vocal Chair Technique

"I was working with a singer in Philly who had a major mental roadblock, which kept him from going deeper within himself to get a great performance. He would only go half the distance, which was not even close to enough. The vocal performances were so boring, I was falling asleep. After hours of trying to get a good vocal take, with the singer getting frustrated and the producer getting more frustrated, I said, 'I'm gonna go get lunch—I'll be back.' So I left for a while and, as I'm walking up to the control room door after lunch, I hear the vocalist singing like a bird! The same guy that couldn't sing before, all of a sudden was singing beautifully. So I walked in, and I looked into the studio from the control room window, and I saw that the mic had been raised up and the singer was standing on a chair! I asked the producer, 'How did you get him to sing?' He said, 'He's not thinking about the song anymore, is he? He's thinking about trying not to fall off that chair!'"

Figure 4-6. Jonathan Davis from Korn.

Ross Robinson on Digging Into Korn Vocals

"On the third Korn record, the singer, Jonathan Davis, called me in to help with vocals. Basically they were all on speed, and he was drinking Jack and Coke all day long—and he was doing a song about an ex, and he had a picture of her. I put it on the floor in front of him, and I got on his back, and I told him, 'Anytime I don't feel you giving, I'm going to hurt you.' And I did that by digging my nails into his skin. Y'know, it was because I was so frustrated probably with the fact that I thought the tracks felt—well, really lame. And also that they were on a click track. When I worked with them before, they were wild! I wanted it to be like it was in the beginning. It was part of my creation, too, y'know. So I just wanted him to rise above. Obviously, they did a great job for what it was. But it wasn't what we did on the first albums."

Bob Ezrin on Duct-Taping Peter Gabriel

"When we were recording vocals for 'Modern Love' on *Peter Gabriel 1*, he wasn't getting the emotion I was looking for in the chorus when he sang the words 'Aw, the pain! Modern love can be a strain!' After numerous attempts at it, I told him he had three more chances and then he was 'going up the pillar.' In our studio in Toronto, the ceilings were very high and there were two stone pillars in the middle of the room. (It was an old building, and these were structural, but they also added to the sound of the room in a very pleasant way.) Peter wasn't sure what I meant by 'up the pillar' and just kept on singing.

"After two unsuccessful tries, I told him he had one more chance or he was going 'up the pillar.' And, of course, he failed again, so Brian Christian, the engineer, Jimmy Frank, the assistant engineer, and I came out of the control room with a ladder and two rolls of gaffer's tape. Brian, who was built like a linebacker, picked Peter up by his armpits and raised him in the air against the pillar while Jimmy got on a ladder and gaffer taped him to it. Once we were sure he was 'secure,' we let go of him, and he dangled up there from the sheath of gaffer's tape around his chest, and I said, 'Mic him.' That's when he sang the version that is on the record."

Figure 4-7. Peter Gabriel's first solo album.

LIGHTS, CAMERA, VOCALS!

Sometimes you just have to give the singer a role and a part to play. I was working with a brilliant Norwegian band named Animal Alpha but could not get the singer, Agnete, to give me the vocal performance I had seen from her when I saw the band play live, months earlier. Onstage she was a wild animal, barking and howling, wearing a wedding dress, smoking a cigar, and spitting into the crowd. In the studio, she was a timid little creature wearing a nice blouse with barrettes in her blonde hair. I wanted to conjure back that stage persona, so I suggested she get dressed in her stage outfit so we could shoot a video of her doing vocals in the studio. I told her it was specifically for a video and that we would not actually be recording the audio. I lied. This was the trick that worked. She became that alter-ego stage vixen character for the camera and gave a dazzling performance—and we didn't even have the camera running. But what a hella great vocal for the song "Bundy" we recorded that day!

Figure 4-8. Agnete Kjølsrud, singer from the group Animal Alpha.

NAKED AND VOCAL ABOUT IT

Some singers insist on recording naked. It is a freeing thing that empowers them to reach beyond their normal range. For this reason, creating a vocal space where a singer can do whatever they want, including stripping down, may be required. Cover the windows, and have soft mood lighting, comfortable upholstered chairs, and places to hang clothes—this can apply to more than just singers. I worked with a Salt Lake City band whose bass player insisted on being naked during the whole session. In fact, the whole band spent most of the session naked. I, however, kept my clothes on.

Figure 4-9. Recording in the buff . . .

Ed Stasium on Putting Mick Jagger "Onstage"

"I'm one to let the artist do their thing. With Mick Jagger, he couldn't get the vocals on this one song—called 'Soul City' originally, renamed 'Peace for the Wicked' later—for his solo album. He knows what he's doing. He goes out there and sings and knows if it's a take or not. You don't need to give him guidance—he's fucking Jagger! Are you kidding me? But he couldn't get the vibe on this one song, so we set up a couple stage monitors in the live room, and we put him out there with an SM57 in his hand. No headphones. He was 'Mick Jagger' out in the room at Right Track, just dancing, doing his stage thing. One take and that was it!"

Figure 4-10. Mick Jagger with Eddie Gorodetsky (L) and Ed Stasium at Right Track in New York during the recording of Peter Wolf's *Lights Out*.

Linda Perry on Sierra Swan's Straitjacket

"One time, when I was working on a Sierra Swan record with Bill Bottrell, Sierra just could not get out of her own head! And the song she was singing was all about being insecure. So I asked myself, 'What would make her feel really insecure right now, to fuck with her?' So I sent my guys to go get a bunch of cameras with flashes, and I put her on a pile of amps, with the vocal microphone all the way up there. And then I put her in a straitjacket. I knew I wasn't going to get a vocal take with all this, but it was just really to fuck with her. Then the whole time she was singing up on the amps with the straitjacket, I had all the guys taking pictures of her. She's just feeling weird with all these cameras going off. She couldn't even mentally process it anymore. I could tell she was very uncomfortable, and that's when she got into character! Then I said, 'Now we can go do your vocal.' I told everyone to go away, put her back in the booth, and—BAM! Got the take!"

Shelly Yakus on Stevie Nicks's Half-a-Smile

Figure 4-11. Stevie Nicks.

"I worked with Stevie Nicks, and she'd have dozens of her closest friends in the sessions, a lot of hangers-on. There was always some turmoil going on because of the people. But for Stevie, her life was the studio and her music, and still is. So she would walk up to the mic, and I'd say to myself, 'She's going to do it again,' and I'd punch Jimmy (Iovine) and say, 'Watch, she's going to do it.' And she'd walk up to the mic, get this kind of 'half-a-smile' on her face, and start to sing with the track. And I don't even think she knew where she was for three and a half minutes. She went somewhere—so where'd she go? Did she go to that place where she was when she wrote that song? Did she go to a safe place where there's no CNN and there's no ex-boyfriends, there's no record label, there's no pressure? Where does she go?

"And when she was done, all of a sudden she'd be right back in the room with her friends and whatever was going on before she started singing—you know, the argument with the producer. But for those three and a half minutes, she was somewhere else. There are some singers who naturally go to that place. Wherever that is. It's not like these people are just taking the train to the next stop—these people are going into the deepest space they can go."

Jay Baumgardner on Studio Posses

"I've partied through entire records. Like, literally. For a while, it seemed every session was a big party, with all the band's friends coming to visit—Papa Roach and other records of that era in particular. I was able to focus on what I did and not let external things bother me. I have a strong constitution and an easy personality like that. Occasionally, there's some overzealous friends of the band—'No, dude, try this idea.' And sometimes the band's friends have good ideas, as crazy as it seems."

GO OVER THE TOP

Often, singers will mumble their words, making it hard to understand the lyrics. To cure this, I ask a singer to give me a full take of the song with them over-exaggerating the words—opening their mouth wider, making a round "O" with their lips, baring their teeth while they sing. This warms up their face muscles and makes it easy for the performers to express themselves. This is also a way to distract an insecure singer from feeling their vocal chords are about to give out.

I find the shape of a singer's mouth will control the brightness of their voice. Often, I will ask the singer to put a smile on their face while they sing. Even if the lyrical content of the song is dark and moody, the shape of the vocalist's mouth into a smile will brighten and add clarity to the voice without making the performance "happy"-sounding. It is like an EQ on the vocal without adding any processing at all.

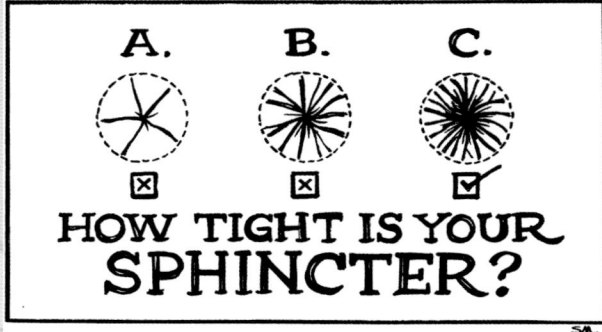

Figure 4-12.

A PUNCH IN THE GUT

I grew up in an operatic household. My mother, Ida (Kuzava Massy Lorinczi) Moody, sang lead in many professional productions, and as a baby, I was in the audience, loving every minute of it—those booming voices, exaggerated movements, and huge bright choruses filling the theater. At home, she rehearsed every day and held me as she sang. I didn't realize until now that holding me was part of her vocal technique. Here is what she says about being a singer:

- "Stand like you are getting ready to take a punch in the abdomen. Like a fighter's stance. One foot slightly in front of the other. Rotate your pelvis under and clench your sphincter! This will give you diaphragm strength to really control your breath for both the loud and delicate parts."
- "Know the score, literally. Know the entire opera, not just your part. You need to know the entire story in order to be the best at your particular part."
- "When singing, feel the sensation of the notes resonating through your sinuses and up through the

Figure 4-13. Ida (Kuzava Massy Lorinczi) Moody.

top of your head. You will be surprised at the high notes you will be able to reach."
- "Hold out your hands in anticipation. The use of your hands is significant—they convey emotion. They will help express the song's meaning."
- "In order for you to be the best, you need to think that you *are* the best. Most importantly, you need to find joy in what you do."

Ross Hogarth on Sucking

"In musicians, there is such a fine line between 'I am the shit' and 'I am shit.' With guitar players, there is a delay—you get a few more takes between the 'I am the shit' and when their brain switches to telling them that they suck. You can watch the diminishing returns occur. With vocalists, it's instantaneous, because the brain is connected right to the vocal. So the minute they go to that place of 'I am shit,' that take is going to suck."

Figure 4-14. Ross Hogarth.

Linda Perry on Christina Aguilera's Ad-Libs

"I work with Christina Aguilera, and talk about ad-lib crazy! I'm telling you, that girl—she's pretty impressive. Because she wants to do her ad-libs right away after her main vocals, and she knows exactly what she's going to do. They are really well thought out. She'll say, 'OK, I want this one to go right between this word and this word—' And she pretty much doesn't let me keep anything she doesn't like. There are no other takes. However, there have been takes I've kept a couple times, because I just thought she was wrong—and luckily I did, because we wound up using the ones that I saved. But Christina will just come in and sing these ad-libs, and she'll do the same thing ten different times, and it sounds exactly the same every time to me. And then I'll catch the one she's trying to get perfect. Then I'll hear the tiny difference—just a teeny little difference that she wanted in there. Just that microscopic inflection of that third note in that run was a little bit out. Her ears are insane. That girl can pick up anything!"

Figure 4-15. Christina Aguilera.

PERFORMANCE OVER PERFECTION

Technically perfect vocals are nice, and in certain cases when harmonies depend on pitch and timing to hold together, they are essential. But the nature of singing is to convey emotion. Emotion is not perfection. When conveying a story about heartbreak, I want to hear that vulnerability in the singer's voice. This doesn't happen when the vocalist is concentrating on pitch and timing. They need to know the song, know what it means, and then they need to sing the song as if it is their own story, being told for the first time right there in front of you.

Heartbreak should not be perfect—in fact, it is a broken human condition. The vocals conveying the story of heartbreak should also be broken. At least, when I record vocals, I want to have at least two takes in which perfection is not considered at all. Just let the singer wreck themselves, break down, completely absorb themselves into the character in the song. If it is their own lyric, have them bring back the thoughts of that moment, when the story was fresh. The best heartbreak-song performances are those when you can really hear the singer cry. It is real. It comes across.

If the song is not their own, the singer needs to channel the energy from the song story and make it their own—become an actor, playing the part of the heartbroken soul. Fixing timing and tuning problems is easy in digital recording. Always go for performance over perfection.

Al Schmitt on George Benson's First Takes

"When we did the *Breezin'* album with George Benson, we did eight songs. When we went to record 'This Masquerade,' I had everybody sitting around the drummer—sitting around the middle of the room here in Capitol A—and George was on a riser with his guitar right there. I looked around to grab a mic and there was this EV666 (Electro-Voice). I just grabbed the thing, putting it on his vocal, figuring we'd overdub it later. Well, he killed it! First take! Wow! That was the vocal you hear on 'This Masquerade'! That was done on an EV666, and the funny part of it was, when I did the next album with George and I wanted to use a U 47 on his vocal, he said, 'No, no, no! Gimme that old gray mic.' It took me a while to convince him that we could use another mic on his vocal. We probably could have gotten a better vocal sound, but we certainly wouldn't have gotten anything more emotional."

Figure 4-16. George Benson's *Breezin'* album.

Matt Wallace on the Essence of Music

"Beat Detective–type programs for drums and Auto-Tune–type programs for vocals take away a drummer and singer's style. For instance, how a singer gets to a note—they might start flat and sharp up to it. Sometimes that is the moment where the singer's real character lives. Or, how a drummer might lay the snare back behind the beat, or move his hi-hats up a bit in the groove—that's their style! If someone ever Pro Tooled Frank Sinatra or John Bonham, everyone would say, 'What happened?' When you put things in perfect time and perfect pitch, you lose what's essential in the music. It's that ebb and flow, that tension and release, that really makes great music."

Al Schmitt on Frank Sinatra and Perfection

"Unfortunately, because we can do all these things to manipulate the recordings now, we are fixing every mistake and everyone is singing perfectly, and sometimes we make records that are just too perfect. We take the emotion out of them. I mean, nobody ever had to tune Frank Sinatra. His pitch wasn't perfect every time, and y'know, if there was a little mistake on something, it added to the emotional effect of the record.

"If somebody in the band played a wrong note, but the emotion was right there, you kept that take, and that was the one that would come out. You didn't keep going over and over and fixing everything to make it perfect. And there was a charm about those things. And I was blessed working with Sam Cooke and Sinatra and Rosemary Clooney and Billy Epstein. People who were just great singers! They sang in tune 99 percent of the time and knew what they were singing! They knew about the lyrics; they knew what the song was about."

Figure 4-17. Frank Sinatra in session.

THE TROUBLE WITH HARMONY

The moment you find your singer struggling to match a harmony to a lead vocal track, turn off all the other parts and have them learn to sing the harmony part as if it is the "new" lead part. This is a trick I learned while watching vocalists struggle to record already written harmony tracks. It is actually not necessary for the singer to listen to the original part to harmonize with it. Suddenly, it doesn't matter if they match the original vocal. And you will be surprised that, when you turn all the tracks on after recording this way, more likely than not, the harmony tunings and timings match perfectly.

Figure 4-18. The Bee Gees.

Ed Stasium on the Bee Gees' Great Big Harmonies

"Watching the Bee Gees doing their backing vocals was a remarkable thing. It was up at Le Studio Morin Heights while I was on staff there. They were singing 'You Should Be Dancing' for the *Saturday Night Fever* soundtrack. They were so precise with their backing vocals. They would all sing the same part, all three of them at the same time on one mic. Then they would double-track that part, then bounce it down to one track, then they'd work on the next part in unison, double that, bounce that, and so on. They were never singing three-part harmonies at the same time. They would work on just one part of the harmony in unison at a time, making these layers on one chorus in the song at a time."

AND NOW HE'S SICK, BUT OF COURSE

The singer's instrument is a sensitive one, and is affected by the singer's own thoughts to the point of complete disability. From my observations, around 80 percent of vocalists are sick in the days leading up to their scheduled vocal recording sessions. They are either deathly ill, or are suffering some mystery malady in the throat right up to the point that they start singing—and then, miraculously, they snap out of it and they are fine. So don't be fooled by a singer's precarious health. Get them in the vocal booth and get them a couple run-throughs.

Nick Launay on Tim Finn Breaking Through

"I did this record once with a great singer, Tim Finn from Crowded House. He would do a guide vocal, and it was always amazing. Come time to do the actual vocal, he got sick. And he remained sick. 'I can't do the vocal, I can't do it!' I thought, *How am I going to get around this?* I thought of which song would sound good with a growly, growly, sick voice and I told him that was the song we were going to do. And he sang it, growly and all, and he sounded sick but perfect for the song—actually, it sounded really good. I played it back and he said, 'Yeah, that does sound really good.' By the third take, he didn't have a cold anymore. A lot of this 'sick' stuff was just in his mind."

Figure 4-19. Tim Finn.

Elliot Scheiner on Lazy Vocalists

"I'm not going to say who it was, but I worked for this band out here a while back, and their lead singer would come in at six in the evening, and we'd do ten or fifteen passes, and then we'd comp a vocal and listen to it. We'd know that we didn't have this or that, and then we'd go back in and punch in those sections and comp them together again. It worked out great, but that was right before Pro Tools. Then when Pro Tools came out, I did another record with these same guys, and the singer came in at ten in the morning this time and did two passes and said, 'That should work, right?'"

Susan Rogers on "Being on Display"

"After working with Prince for many years, the first band that I worked with was the Jacksons, and Jackie was a particularly reluctant singer. He didn't have the same voice that Michael and Jermaine did, and he was uncomfortable doing vocals. So I showed him the Prince technique of how to record vocals by himself. He described to me what it was like being a kid at Motown, that when the Jackson 5 would do their vocals, the secretaries would show up to the sessions and they would bring their husbands and their children—the control room would be packed. These Jackson kids felt so on display all the time. I'm glad I was able to introduce to him the process of 'being by yourself.' I've talked to other singers about that. Most of them don't want to be by themselves—they like having an engineer or others in the control room."

If they seriously have a problem, you'll hear it. But getting started may be the only way for them to figure out that they can actually sing and everything is OK. Poor singers. So delicate. So sensitive. So spoiled! Kick their frickin' asses!

THE POSSE IS IN THE HOUSE

If you want to get a singer out of a funk, try bringing in an audience—suddenly, they are performing as if they are onstage! Even if there are just three or four people in the control room that they don't know, this will rattle them out of the "I can't sing today" moment in a hurry. Sometimes a reluctant singer just needs someone to sing to. I often tell singers to imagine themselves in front of an audience. When I worked out of RadioStar's theater, on occasion I would set the vocal mic up onstage so the singer could project out over the auditorium, having them sing to the empty seats. This made it easier for them to imagine being onstage at a show.

Figure 4-20. Simple vocal aids.

REMEDIES WITHOUT PHARMACIES

A few things I have on hand to make vocal sessions go smoother: Throat Coat tea, Breathe Right strips, and a neti pot. The tea is a special blend made with slippery elm and licorice, which actually helps keep the singer's throat moist, relaxed, and calm; with milk and honey, it is a comfort on every vocal session. The Breathe Right is a simple, stiff Band-Aid-type adhesive strip that comfortably props open the nasal passages. And the neti pot will clean out the sinuses with simple salt water. Pedia-Pops can also help a singer who may have stayed up partying too late the night before—they gently replace electrolytes and rehydrate the poor babies!

Shelly Yakus on Singer Lung Power

"Most of the great singers I've worked with do a lot of cardio workout. They'll run, or they'll go to the gym and do the treadmill. They'll do forty-five minutes a day, and they get real lung power because they've exercised. Go to the gym. Run or jump rope or whatever. You'll be able to get more air in your lungs. Then when you sing, you can hold the ends of the words out in tune, because you have enough air control."

POP-UP VOCAL BOOTH

When I recorded System Of A Down at Rick Rubin's home studio, I had to record in a big cement room full of creepy antique taxidermy mounts. There was no dedicated isolation booth. Not a great place to do vocals. My solution was to buy a large camping tent, tall enough to stand up in, and that became our vocal booth. Serj Tankian, the singer for SOAD, set up the space as his private sanctuary. This helped him to deliver brilliant performances. I have yet to get the pot smell out of that tent.

SYLVIA'S SECRET VOCAL COMPRESSION

OK, I'm giving it up now: my magic combination for capturing exciting and dynamic vocal performances. First I'll use a Telefunken U 47, patched directly into a Neve 1073 mic pre/EQ set only one click up on the mic pre. Then I patch out of that into a soft tube compressor—preferably a UA 175b or a Sta-Level. (Reproductions of these units work well also, if originals are not available.) Then out of that compressor into a second compressor! I prefer a Urei 1176LN, but an Empirical Labs Distressor is also an option. Then I use light EQ, bringing out the mid-range on the vocal for clarity. Just a touch.

I set the compression by having the singer, or preferably another voice, give me the dynamic range. First, I'll have a vocalist sing softly on their own, without the track running, singing a verse part a capella. I will adjust the tube compressor to react to this lower level, without engaging the second compressor. Then I'll have the singer get loud. While the vocalist is loud, I'll adjust the second compressor to react to the loud voice, limiting the

Figure 4-21. System Of A Down's Serj Tankian.

Figure 4-22. Sylvia and her true love.

loudest passages. I will actually ask the singer to get really loud—louder than they will be singing—and here is why: Singers are generally shy about getting loud. Once the track is rolling and they are singing their parts, they will automatically get louder. You will be disappointed if you are not ready for that first take. You may even get fired. I developed this vocal compression setting to protect myself from distortion on that first take. This is the basic technique I've used for singers from Billy Corgan to Johnny Cash to Serj Tankian.

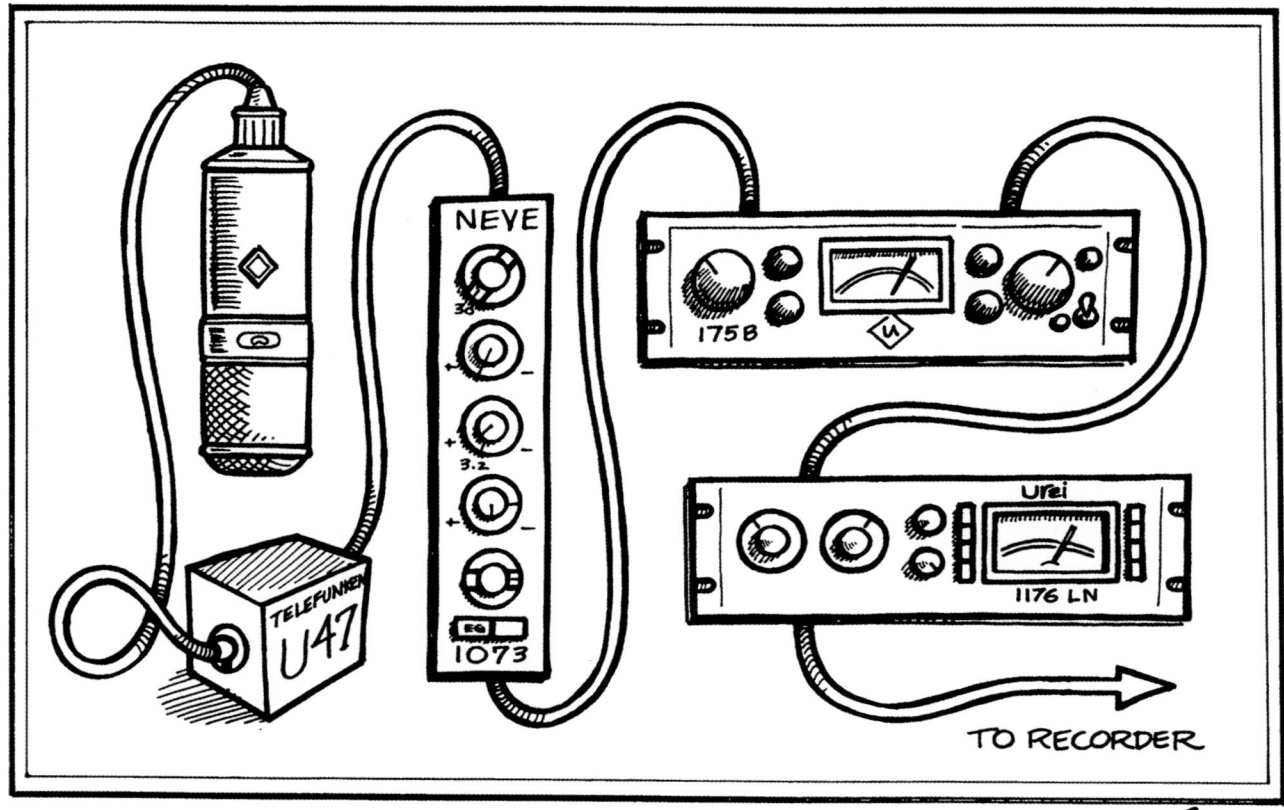

Figure 4-23. Sylvia's Secret Vocal Chain.

HEADPHONE MIX TO CONTROL VOCAL PITCH

You can manipulate the pitch of a singer by controlling the mix and levels going to their headphones. When a singer is out of tune with the track, try adjusting up the elements in the headphones that the singer can cue off of. For instance, turn up the piano or guitar. The bass guitar may be too low of an instrument for this to work, and other vocal tracks could cause confusion.

Another extremely useful technique is to have a singer pull the headphones off of one of their ears. This will allow them to hear the track for pitch reference in one ear, while hearing their own voice in the room with the other ear. When a singer is consistently sharp against the track, this trick is often the easy fix. Another helpful technique is to adjust the singer's level up or down within the headphone mix. Many times it is much better for the engineer to make these adjustments than the vocalist themselves, so be sure that if you are engineering, you can hear exactly what the singer is hearing, to help them.

Linda Perry on Vocal Levels

"If a singer is going way out of tune, I will go into the vocal booth, and I'll put on their headphones to see what they're listening to. I usually wind up saying, 'What the fuck?' The vocal will be cranked way up loud in the headphone mix. Most of the time, if a singer has trouble with pitch, it's because their headphones are way too loud and they are singing past the song and the tuning. So I will have them turn it down, and I'll tell them, 'You'll just have to feel this a different way. Stop listening to yourself and start listening to the music!' When they do that, then all of a sudden, they are in the song!"

Figure 4-24. Veruca Salt.

Brad Wood on Headphone Mix for Vocals

"I think headphone mix for me is probably more key than anything else in getting great vocals. It needs to be something that sounds kickass, and I'm always out there checking to make sure. Tomorrow morning, Veruca Salt's vocalists will be in. They usually sing together, and they usually sing standing side by side. Sometimes they are singing in the control room with no headphones. And what they want is it to sound perfect. They come in, headphones go on, and then they go. No waiting. No fooling around for a good headphone mix. I might prep all the songs they would possibly sing on the day before. It has to sound kick ass."

Figure 4-25. Luana Caraffa from Belladonna.

HEADPHONE LEVELS TO CONTROL PERFORMANCE

If you want to get a more intimate vocal performance, try cranking the vocal level in the singer's headphones. This forces them to sing quietly, because their own level is so loud, they need to stay quiet so they can hear the track. Or try the opposite. If you bury the vocals in the track, the vocalist will sing louder to hear themselves above the track. This is an especially good technique for shy singers.

Tim Palmer on David Bowie and Headphones

"Bowie would *never* complain about a can balance, and his vocals were incredible. He is the only singer I have worked with who redid a vocal to make it more melancholy by pitching it very slightly 'flatter.' Amazing."

Figure 4-26. David Bowie.

MODERN MICROPHONE CHAMPIONS

Don't get me wrong—I treasure my Telefunken U 47. But it is often fussy and occasionally has issues. Because of this, I usually lean on other mics to handle vocal duties. There are several great modern manufacturers that have large diaphragm tube mics based off of vintage Neumann U 47s, U 67s, M 49s, and AKG C 12s. Some of these include the David Bock—designed Soundelux microphones and David's own line of Bock microphones. Manley's tube mics are wonderful. Reissue Telefunkens are also fantastic! I am in love with Mojave's MA-200 tube microphone, which is as good as a vintage Neumann U 67, in my opinion. And it never has a "bad day."

Figure 4-27. Mojave MA-200 tube mic.

Matt Wallace on his MXL 990 and O.A.R.

"What could possibly be my favorite vocal mic? Well, my engineer Will Kennedy will confirm this. We would do these blind shootouts where I wouldn't know the name of the microphones—just microphone A, B, C, or D. And I'd put up the best of the best—AKG 414, Neumann U 47s, U 87s—and then I had this MXL 990 mic made in China. You can buy it at Guitar Center for $99. We'd have the singer sing, and we'd pick the best mic out of the lot. And it kept happening. We would pick the MXL 990! There were a couple times where I would say to myself, 'I don't want to keep picking this cheap-ass mic from Guitar Center,' but sure enough, it was the stupid MXL 990 again! I had a couple of friends who were these highfalutin engineer guys, and they'd go, 'Oh, dude, MXL 990? How's it sound?' And I'd say, 'Well, dude, it sounds pretty good on the radio!' I used it on some *big* records! It was on O.A.R.'s 'Shattered (Turn The Car Around),' which was a huge hit. And we did it with that stupid-ass microphone. It just sounded pre-EQ'd. It had the right amount of focus for the vocal. So now I have four of them!"

Figure 4-28. Matt Wallace's MXL 990 microphones.

Ed Stasium on Bruce Swedien's Vocal Mic Choice

"My go-to vocal mic now is the Shure SM7. Go figure! That's what Bruce Swedien used on all the Michael Jackson stuff, so it's certainly good enough for me. Actually Bruce recorded my first and only band, Brandywine, in 1971 at Brunswick's recording studio in Chicago. We did it in three days. Working with him is when I really got the recording bug."

THE ODDBALLS

Documenting a song with a straight vocal recording is one thing—creating a vocal "sound" is a whole other thing. I find there are a zillion ways to add character to your vocal sound, and one surefire way is to use a weird mic. Use a telephone handset, a CB radio, a walkie-talkie, or even a cup and string. Use a really old cast-nickel mic from the '50s. Or even use a guitar as a

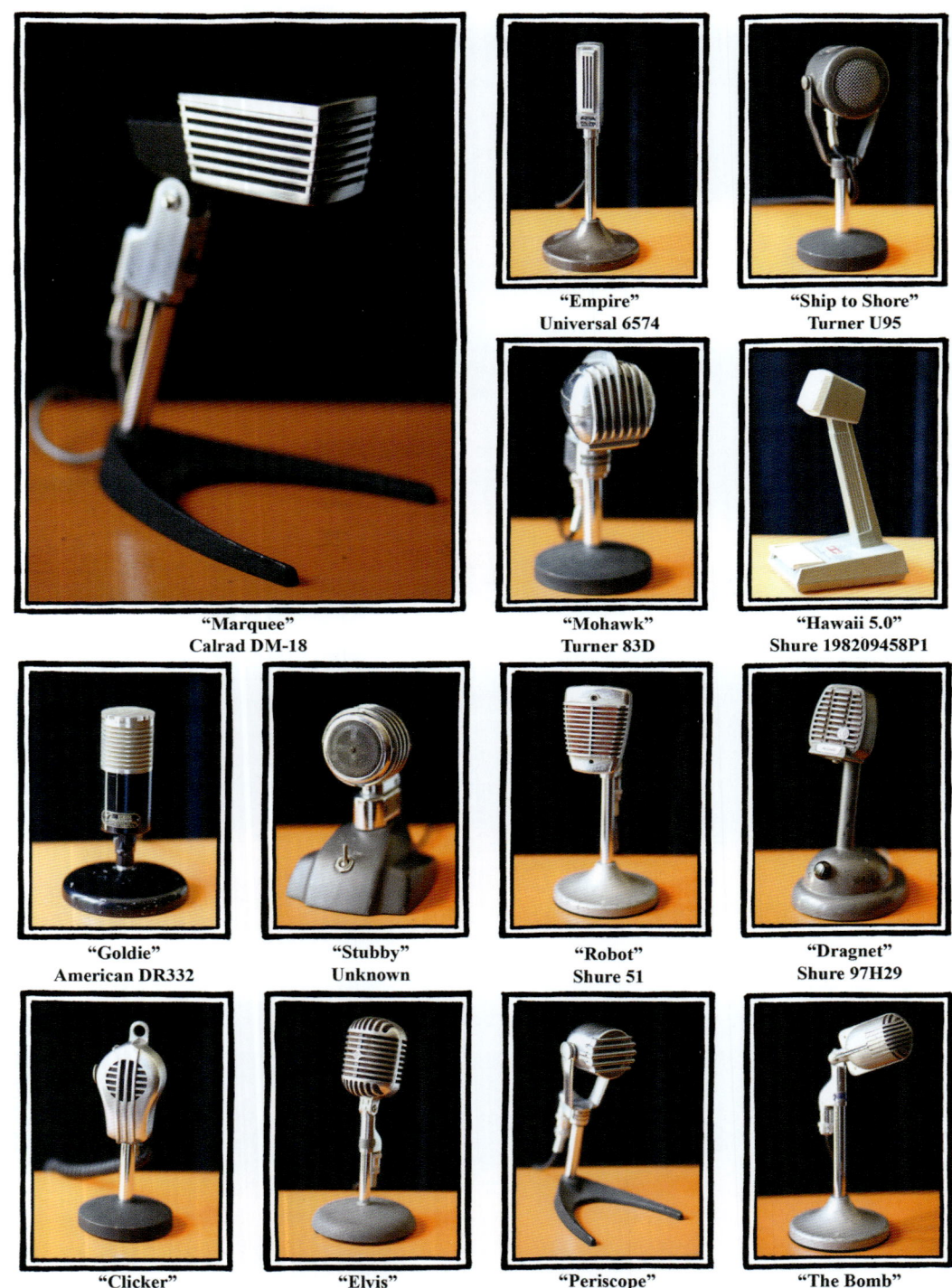

Figure 4-29. Sylvia's Old Mic Gallery.

Mark Rubel on Recording Vocals with an "Ice Cream Cone"

"I love the RCA BK-1A 'Ice Cream Cone' mic; it has a very peaky mid-range. Either by itself or paired with a 'good' microphone, it helps the singer to sit up front."

Figure 4-30. RCA BK-1 microphone

Justin Stanley on Nikka Costa's Strange Little Mic

"I bought an old RCA reel-to-reel from a used shop a while back, and it had a classic little microphone that came with it. On Nikka Costa's first American release, on the song called 'Like a Feather,' I used that mic for all Nikka's vocals—just the mic, not the tape deck. It had a very unique, really great mid-range-y sound."

Figure 4-31. Justin Stanley's little square RCA mic.   Figure 4-32. Artist Parisa Kamankesh records vocals in her bedroom studio.

Figure 4-33. Pressure Zone or "PZM" microphone.

microphone! So many possibilities! All you need is curiosity and a patient vocalist.

MATT WALLACE ON SINGING INTO A WALL

"I use these RadioShack Realistic PZM mics—which, by the way, sound a lot better than the Crown PZMs. These days, the Realistics are getting harder to find. They are 'pressure zone mics' and are known to have fewer phase anomalies because the sound hits the plate and then goes up into the mic. Here is a trick: Instead of putting an AA battery in here, put two 6V batteries in the battery holder. It will take a lot more sound pressure and is cleaner and clearer! I would tape the PZM mics directly to the bathroom walls when I did recording in there. And these are wicked for doing vocals! If you want to get a crazy vocal take, tape a PZM up on the wall and have the singer put his or her nose right up against it and sing. You'll get this most wonderful 'direct'-sounding vocal."

THE COLOR AND THE SHAPE

Sometimes a detailed EQ is what you will need to bring out the best in your vocal recordings. Here are a few EQ choices that can subtly enhance the sound, while having enough control to completely change the character of the vocal when needed.

TO REVERB OR NOT TO REVERB?

Many vocalists insist on drowning themselves in a sea of reverb during vocal tracking in their headphone mix. I acknowledge that this is the standard way of doing things, but I suggest that those cowardly singers are hiding. How can they hear the details of their pitch or performance when obscured behind a cloud of reverb? Try recording without reverb. Get them used to it. Unless the reverb is being used to solicit a certain type of dreamy performance, it should not be necessary.

COMFORT/DISCOMFORT

Controlling the environment during the tracking of vocals can dramatically change the results you get.

Figure 4-34. Rupert Neve Shelford 5051 EQ/Compressor and 5052 Mic Pre/EQ.

Figure 4-35. Linda Perry's EAR stereo EQ.

Larry Crane on Sleater-Kinney, Trust, and Friendship

"Corin Tucker from Sleater-Kinney is one of those singers that will consistently give you a great vocal take—not just in the first few takes. She's a powerhouse. She's got a loud voice and a lot of control with vibrato. She can be really nice while you are just talking, then she'll go in the vocal booth and she flips, and pure rage or anger or whatever she's trying to project is switched on! I don't think I've seen anyone quite like her. There's a song called 'Sympathy,' about her first child. It's real emotional for her. The lyrics were so powerful, it affected everyone in the control room, too. Luckily, we've all known and worked together for so long, I think it was a real safe environment for performances like that to happen. I think the producer, John Goodmanson, and I were more than just hired guns on the sessions—we had all become friends."

Figure 4-36. Carrie Brownstein from Sleater-Kinney.

Ed Stasium on Funny Business with the Tea Party

Figure 4-37. The Tea Party's *Edges of Twilight* album.

"There was a Canadian band called the Tea Party that I recorded, where we built an Arabian tent for [vocalist] Jeff Martin to sing in. Right in the tracking room at A&M Studios in Los Angeles. During the *Edges of Twilight* sessions, Jeff spent hours in that tent with his hookah, mixing hashish and tobacco. Actually, he and his girlfriend were in there together most the time, even when we were recording vocals. I'm pretty sure there was some funny business going on in there during the vocal recording. For God's sake, the mic was open! In fact, on one song during an instrumental section, while the tape was rolling, there was a huge crash and a moan. Turns out those two were doing something in there in the dark and knocked over Jeff's sitar and broke the neck off of it."

Figure 4-38. Pauley Perrette tries to warm up in the vocal booth.

Most producers and engineers will make the vocal space as comfortable as possible, with a chair, mood lighting, candles, tapestries, pillows, lava lamps, etc. I agree that if you are recording a love song or pop song, this is a great way to go; however, if you are recording an angry song with aggressive vocals, I suggest you *not* make the singer comfortable. In fact, turn on the fluorescent lights, chill the room, make the singer sit on the floor between takes. All of this discomfort will be reflected in the performance!

PISSING OFF THE SINGER

Tool's music on *Undertow* was full of blood-vessel-popping screams, like on the song "Crawl Away." I heard those screams onstage, I heard them in rehearsals, but in the studio, Maynard James Keenan's screams were half-jacked and lackluster, even with the perfect complement of the AKG C1000 mic. After several attempts at one of those ten-second screams without a good take, and with his voice obviously wearing thin, I finally asked him to go outside and run around the block five times. This would make him furious, but after he did it, he nailed those screams. He was pissed, and you could hear it in his voice! Sometimes it is better to make the singer as *un*comfortable as possible to get the right performance!

Ross Robinson on Drawing Out Anger

"You're going to find that most anger and resentment is love-based. When I ask questions about that resentment, I get into the core of what the song is about. The angriest vocal performance is the vulnerable one. The one where it's just like 'Fuck it, I'm going to look stupid and I don't care.' And when that heart opens, the most pissed-off, amazing performances happen."

Geoff Emerick on Miking the Back of Lennon's Head

"John didn't like the sound of his voice. And on occasion, he could get really—not stroppy, but sort of irritable. And he could not understand why he had to put the microphone in front of his mouth to record his voice! So I was getting angry at him and I shoved the mic on the back of his neck! Right? And he carried on singing and wanted to hear it! 'Course, when he heard it, he didn't like it!"

Figure 4-39. John Lennon.

HANDHELD VERSUS SUSPENDED MICROPHONES

Because the recording studio can often be intimidating to a singer, sometimes it is necessary to try techniques and equipment not traditionally used for vocal recording, just to keep those singers from feeling self-conscious. That's where the AKG C1000 mic came in on the Tool sessions. It was not my first choice. It was not even on the list of choices, because I had never used it for vocals, ever. I was looking for the sound of a Neumann U 67 on Maynard James Keenan's voice for the record, but because the little troll squats and shouts into the floor when he sings, it was difficult to suspend a Neumann U 67 in the right position for a vocal take. We tried it, spending several hours adjusting the mic so it was a foot off the floor facing upward, to capture Maynard's verbal regurgitation. But ultimately, it failed.

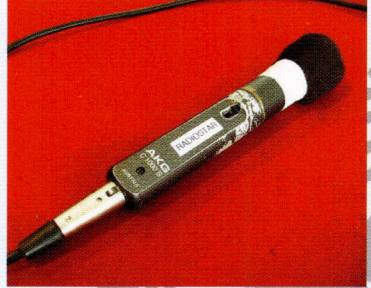

Figure 4-40. The Tool vocal mic, an AKG C1000.

The AKG C1000 was actually Maynard's suggestion. It is not a mic meant to be handheld, but after wrapping it in foam and duct tape to reduce the handling noise, it performed surprisingly well on vocals. Because it was a condenser, it retained most of the sparkle of a U 67, but Maynard could grab it in his fist and smack the mic around. Ultimately, the C1000 was the workhorse on the *Undertow* sessions, but we did use the Neumann U 67 for the more intimate passages with a Urei 1176 compressor cranked way up to expose all the little details in Maynard's voice.

Another technique is to have both a suspended and a handheld mic set on a stand side by side, then record only with the suspended mic. This way, your singer can lean on the handheld. This "dummy mic" technique works great if you don't want to hear the singer slobbering all over the handheld.

Figure 4-41. Maynard James Keenan sings on Tool's *Undertow*.

Figure 4-42. The Devolution of vocals.

Chris Shaw on Flavor Flav's Vocal Mic

"I remember always using a Sennheiser 441 on Flavor Flav during the Public Enemy sessions for *It Takes a Nation of Millions to Hold Us Back* and *Fear of a Black Planet*. He was really rough on microphones, and so we let him hold that one—because it was a dynamic—and he would always slam it with the palm of his hand and drop it on the floor. Plus, it took the edge off his voice. We usually had to spend a bit of time spot-erasing all those noises afterwards, on tape. On an MCI JH24—ouch!"

Figure 4-43. Flavor Flav.

Tim Palmer on David Bowie's Tin Machine Vocals

"I recorded all David Bowie's vocals on the Tin Machine album with a Shure SM57. Still my favorite mic. It was all part of the plan to keep the album from sounding too 'hi-fi.' Who knows why we recorded it digitally! I think that was because they used those 48-track digital machines at Mountain Studios where we started the album. We were aiming for a tougher, more performance-based, less polished production, and I always love the freedom that this mic—or any handheld—gives to a vocal st"

VOCALS 81

Ross Robinson on Saving an Aggressive Singer's Voice

"Sometimes I'll have a crusty mic that I'll put alongside the main vocal mic, which is usually the Stephen Paul modded Telefunken U 47 for me. That way, the singer will monitor off of the crappy mic. I find that when vocals are so clear and great-sounding in the headphones of an aggressive singer, they lose their voices fast. So having the jacked-up crusty mic stops them from trying to push too hard. Their physical voice gets saved, and all the fire comes through. I'm not one of those guys to say, 'You know, you're singing too hard, maybe if you back it off ten percent—.' Who the fuck wants to hear a singer back off ten percent? I don't!"

JUMPING IN THE DEEP END

SINGING IN AN AQUARIUM

The problem with the idea of singing underwater is that singers generally need to breathe. Now, if you have a singer who does not need to breathe, then this could be a fantastic technique to get an otherworldly performance! Or try this: Wrap a mic in plastic and put it into a pail or tub of water, and then have your singer sing into a tube that is submerged underwater. OK, it may be a bunch of bubbles, but it could also actually be pretty cool!

CAUTION ABOUT POWER HAZARDS

Now, when experimenting with mics and water, I suggest using dynamic microphones for a very good reason. Condenser mics have a voltage going to them through the mic preamplifier's phantom power. It may be a low voltage, but it's still nothing to mess with. At the same

Ron St. Germain on Bathtub Vocals

"While recording the Canadian band Joydrop at Longview Farm, we filled up a bathtub in one of the studio's apartments so we could record vocals on the song 'Until.' We wanted the sound of 'bubbles' in the vocals, so I had singer Tara Slone actually get in the tub and submerge briefly underwater while I recorded her from just above the surface. Although a slow process (due to the whole 'breathing' issue), I found it a really unique sound that I have yet to find on any effect processor."

Susan Rogers on Underwater Westerns

"Tommy Jordan wanted to make music for an underwater western—a whole western movie underwater with his band Geggy Tah, and I asked him, 'Why?' He said, 'Because that's the next stage west.' Well, OK, we are on the West Coast, and you can't go any further west unless you go underwater! So, yes, it clearly made sense!"

Figure 4-44. Dani Aubert from Patchy Sanders tries something different.

time, tube microphones often are connected to power supplies that carry quite a bit more electrocution power—so no tube mics in the bathtub, please!

SINGING THROUGH A SNARE DRUM

This is a technique I've seen Steve Albini do, though he may actually be singing "to" a snare drum. I tried it on the Patchy Sanders album, and it gave the lead vocal a halo of a papery harmonic. Very interesting.

SINGING THROUGH A FAN

All right, as if the electricity weren't enough, try singing through a fan for a *fan*tastic sound on your vocals (or any other instrument, for that matter). The quieter the fan motor, the better. Singing into the fan while miking its intake side will reduce wind noise; or try a windscreen on the mic if having the fan blowing in the singer's face is too much. And I suggest keeping the cover on the fan to keep your singer's nose from being compromised.

Geoff Emerick on the Beatles' Leslie Vocals

"On 'Tomorrow Never Knows,' John Lennon wanted his vocal to sound like the Dalai Lama on a mountaintop twenty-five miles away, and by chance I'm looking out the control room window, and I see the spinning speaker on the Leslie. So, I'm thinking if I could break into the circuitry of that and put John's voice in there, we might give him an answer, y'know. I fed the vocal through the desk and out one of our foldbacks, or whatever it was. I know we had problems with feedback and all kinds of stuff between the microphone and the speaker, but we got it there. It had never been done before, and it was by chance and by luck that it happened, because John had these ideas, but he didn't know how you would ever make them happen. None of the Beatles were technical whatsoever."

SINGING UPSIDE DOWN

Poor Serj. I had him hang from an exercise bar in Rick Rubin's basement, holding an SM58, to try the technique of singing upside down. When he screamed a vocal part for System Of A Down's debut album, I thought his eyes were going to pop out! The take was not usable, and I quickly abandoned the idea of hanging a singer upside-down.

Well, there may actually be a positive result from having a singer hang upside-down. Apparently, when a singer leans over, putting their head between their knees, you reverse their perception of head and chest voice. It then becomes much easier for a singer to move between head and chest voice without the strain. The technique was first described by author Eloise Ristad in her book *A Soprano on Her Head* as a way to train a singer to cross ranges without "pushing."

There may be a lot of wisdom in this technique. While recording vocals on Tool's *Undertow*, Maynard James Keenan insisted on leaning over and facing the floor while singing, so he could control his range without pushing—and it worked! I adjusted the Neumann U 67 mic so that it faced upward, only inches from the floor, so he could face down into it while singing.

Figure 4-45. Serj Tankian, singer from System Of A Down.

Ross Hogarth on Couch Singing

"I dealt with a singer that said, 'Hey, I want to sing while I lie on my back on the couch.' I said, 'OK.' So we did it. And when they were on the couch and singing, they had no air in their diaphragm. They sounded like they were singing on the couch. Then I brought them into the control room and played it for them. I had to accept that if they liked that, I would keep going, 'Look, my name is not going on the record as the singer, so if you like this sound, I'm happy to continue down this path.' As an engineer, you have to be ready to do that."

Brad Wood on Multi-Miking of Vocals

"Especially loud singers I'll put through this little Dukane PA amp and drive it hard. I'll be recording the mic they are singing into, plus a second mic that is picking up the vocals going through the little Dukane amp and speaker, plus a third room mic that is picking up everything."

Figure 4-46. Brad Wood's Dukane amp.

Bruce Swedien on Michael Jackson's Tube Vocals

Figure 4-47. Michael Jackson's *Thriller* album.

"On one occasion, I recorded Michael singing into the vocal microphone through a four-foot-long cardboard mailing tube. I attached the tube to a mic stand and aimed the tube into a microphone with Michael singing into the open end. The song was 'Billie Jean.' I wanted to have a very unusual effect as part of the sound field of this great song. It occurs at the end of the third verse; he's singing, 'Do think twice.' It has a totally unique sound quality. Not electronic at all. Very different! Another time I recorded Michael singing into a mic in the tiled bathroom at Westlake Audio in Los Angeles. The space had a highly reverberant, short decay sound quality. It was great!"

Larry Crane on Transgender Vocalists

"I recorded a rock opera for an actor/singer guy. On one song, the singer was a person who was going through gender transition from female to male. This is the most odd thing to happen to me, but he sang a part and came in to listen and said, 'I hate it. I sound like a girl.' I wasn't quite sure what to say—luckily, his friend came in and they started laughing about it. Whew, got me off the hook."

HOW TO BUILD A TELEPHONE MIC

First, find an old retired landline-style phone. It will need a handset with a carbon button-type microphone element inside the mouthpiece. I like using the classic dial-type phones because the handsets are easy to take apart. Take a sacrificial XLR cable and cut the female end off. Pull the ground wire aside on the cable and separate the two leads. Unscrew the tele-

phone mouthpiece and pull out the carbon button's holder and remove all the old wiring and the curly telephone cord. Next, wire in a 1.5-volt battery across one side of the XLR leads. I just solder it right in there. Connect the leads from the XLR to the two posts on the carbon mic's holder and you are ready to go! Stuff everything into the handset, pull the XLR cable through the opening, and screw the cap back on. The signal from the handset will be line level, and remember, you are not going to hear anything out of the earpiece. But when you plug it in, you'll get a microphone that will sound just like a telephone! My father, Don Massy, taught me this simple and fun microphone project.

Figure 4-48. Don Massy's telephone mic.

HAVE A DRINK ON ME

I once worked with an Australian singer who failed when it came to vocals. Just sucked. I couldn't figure out why, because he had been so good in rehearsals. Then I realized that in the rehearsals he always had a bottle of vodka and in the session he was stone-cold sober. I immediately went to the store and bought him a quart of Absolut. Vocals after that were brilliant! Lesson learned: Quit drinking after vocals are finished!

Figure 4-49. Don Massy and baby Sylvia.

Ross Robinson on Dropping Vocal Pitch

"I was working with this band from Slovenia named Siddharta, and the singer would sit on the balcony and smoke and drink for three days straight before his session. Drinking beer and smoking cigarettes. And his voice would drop, like almost an octave. It was the most amazing technique I've ever seen in my life. And he makes a plan to do this every time, so his voice will drop low for his vocal sessions."

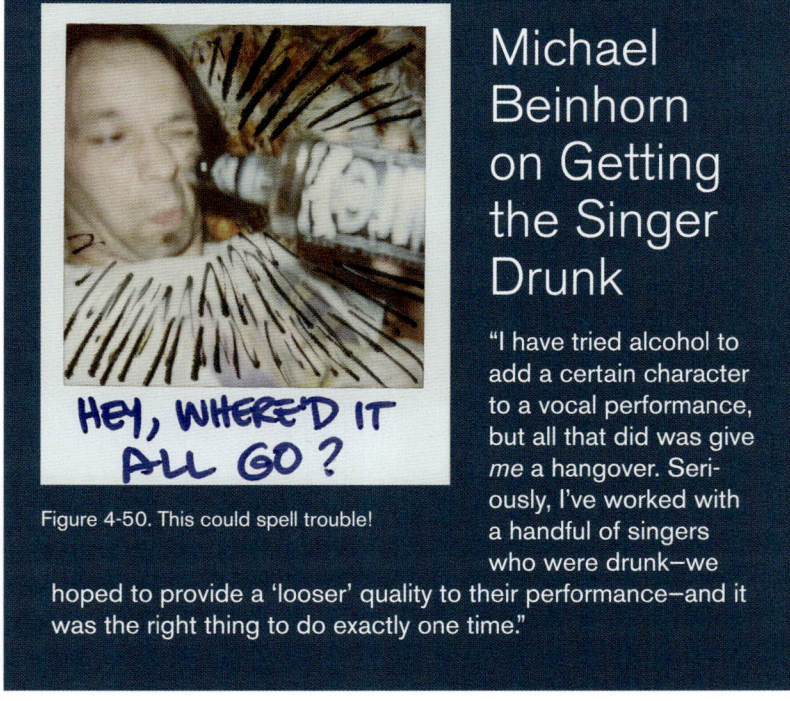

Figure 4-50. This could spell trouble!

Michael Beinhorn on Getting the Singer Drunk

"I have tried alcohol to add a certain character to a vocal performance, but all that did was give *me* a hangover. Seriously, I've worked with a handful of singers who were drunk—we hoped to provide a 'looser' quality to their performance—and it was the right thing to do exactly one time."

Linda Perry on Bad Vocal Advice

"When 4 Non Blondes did our first record, everybody was on my case because I smoked like crazy and I drank like crazy. Everyone was saying, 'You can't do that when we go make the record. You've got to get vocal lessons, and you've got to be strong when we get in the studio.' The manager was telling me this, the producer was telling me this. I was green and I didn't know shit, so I said, 'All right, I'll do it for you guys.' I'm not going to be an asshole; I'm going to listen. So I stopped drinking, I stopped smoking, even stopped drinking coffee—got up to the microphone to sing, and I was like, 'What the fuck just happened?' I had this high, crystal-clear fucking voice! 'Who is that? That's not me!' Mind you, it was strong, clear, and it was high and it was great! But it was not *me*. I'm dirty and have a gritty, low-register voice. I'm not trying to run around winning honors here. I don't need a gold star for anything. So I said, 'Fuck that!' And I immediately started smoking and drinking."

Figure 4-51. Linda Perry and Sylvia at Linda's Kung Fu Gardens.

Matt Wallace on Staying High

"For me as a producer, sometimes you have to keep the booze and the drugs away from people and sometimes you have to absolutely feed it to them. It's true. And sometimes there are musicians that say, 'Dude, I can't believe you're making me smoke weed and drink whiskey right now!' And I respond, 'Yeah, well, it's because you sing a lot better with it!' Just because you are going into the studio doesn't mean you should stop doing what you normally do. Just smoke some weed and sing the damn song!"

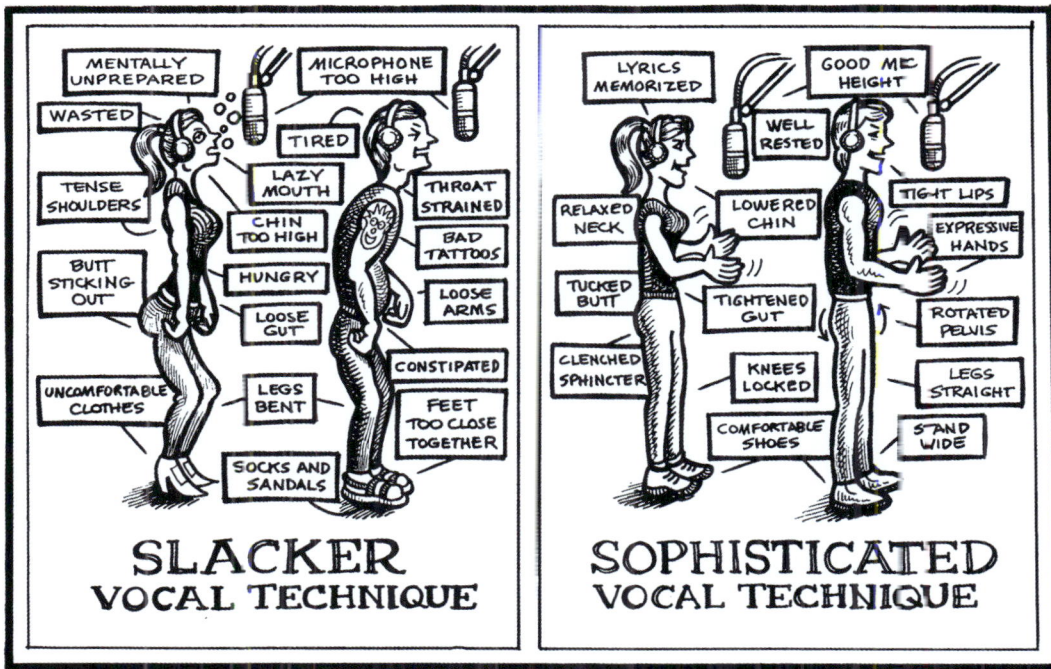

Figure 4-52. Slacker versus sophistication.

Figure 5-1.

5

BASS

Figure 5-2. Filipe from Klepht laying down bass with a glove.

FINGERS, PICKS, AND WHAT?

Fingers—the bass player's default mode—allow for optimal dexterity. Using a pick—ooh, this guy's got skillz, with clarity and consistency. But playing bass with gloves on your picking hand—wait, what?

Don't laugh, because it actually worked out pretty well on the Klepht album. Bass player Filipe Content wore gloves to give the bass a soft, boomy tone.

SEXY, SEXY BASS

A trick to get the bass player in the groove is to have the player play his parts slightly behind the rest of the band's tempo. This works for instant pocket. If your bass player can't quite lay back enough to find the sweet spot, you can always find it for him after the part is recorded, by moving the digital track later by 10 to 20 milliseconds. This works for guitar, too. Damn guitarists and bassists, always rushing! Gotta do it like the bass master Bootsy Collins! It's Bootsy, baby!

THE MAGIC COMBINATIONS

When anyone asks me to recommend a four-string bass/amp combination, it is always Fender/Ampeg—more specifically, a Fender Jazz Bass with an Ampeg SVT head (preferably vintage 1979). Why? Because the bass in your recording will have character. With that rig, you can get a big growly tone for rock music, and if you back off the drive, it will remain sweet and deep, with every string heard.

If you need a five-string, a Warwick Corvette is the instrument. Try it with a fabulous Krank rack-mount amp head. On the Tool records I recorded, bassist Paul D'Amour used a Chris Squire model Rickenbacker 4001 bass guitar with a Mesa 400+ bass amp head, driving an Ampeg 8x10 bass speaker cab.

Figure 5-3. Cecil Gregory poses with the biggest, baddest Fender/Ampeg bass rig.

Figure 5-4. Paul D'Amour from Tool with his Chris Squire Rickenbacker 4001.

Fancy-Schmancy DI Boxes for Bass

This may be even more important than what mic you put on the bass speaker—the DI or direct box. It is an easy way to go directly into the console without using a speaker at all (but I would never recommend using a DI on its own, unless you were trying to reduce bass bleed in tracking). I've tried a zillion makes and models of DI boxes, and some of them are very expensive and fancy, but I think the rugged Countryman DT-85 DI is the best and most consistent of all DI bricks. And they cost a few hundred bucks. Go ahead and spend thousands on tube mic pres, if that's what you want—there are some great ones out there. They might be good for many things, but you can always depend on these little Countryman black boxes.

Figure 5-5. Eric Wilson from Sublime knows the importance of a good hat.

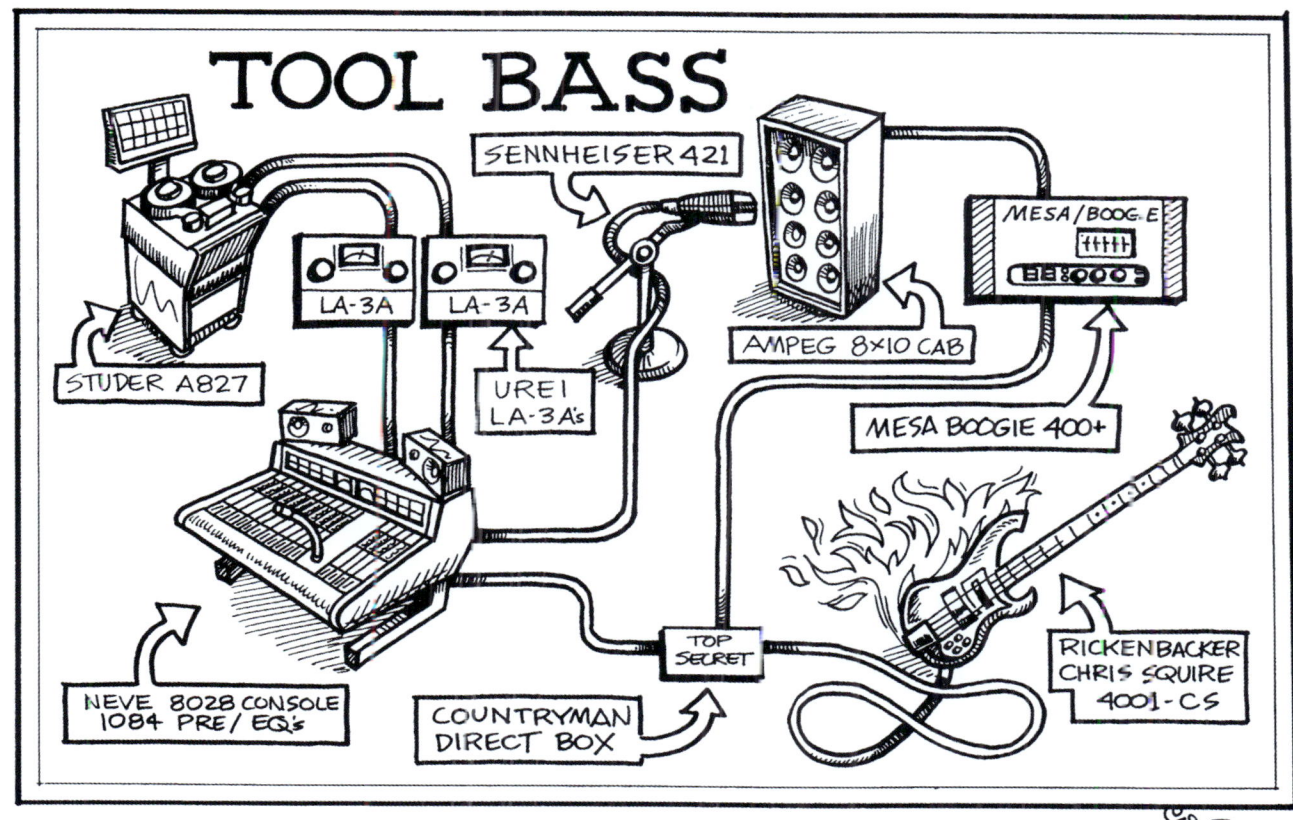

Figure 5-6. Sylvia's Tool bass-recording chain.

Figure 5-7. Girls like big mics.

The Rickenbacker basses have a very distinctive tone, and you can hear its mid-range-y character on Yes's *Fragile* album.

LOW-DOWN MICROFUNKEN AND EQ

You can stress out about the sound of the bass all day long, or you can just throw a trusty Electro-Voice RE-20 on the cabinet and get a great sound instantly. It's a big mic for a big sound. That's right—I'm a woman and I like big microphones. You figure it out.

Then patch that signal into a Warm Audio EQP-WA. Kinda looks and acts like a Pultec, for a tenth of the price. Sign me up! Sign me up twice!

Figure 5-8. Warm Audio's EQP-WA. (Photo courtesy Warm Audio)

BASS 91

ABSOLUTE DISTORTION ON EVERYTHING

There is something very exciting about distortion on bass—it livens up what once was just a low, dull pulse in music, makes it stand out and talk like the other instruments. Distortion in general is a wonderful thing! Especially amplifier distortion through overdriving the input. It can create an edge or an outline to the tone. Other ways to get "quality" distortion include the Sans-Amp distortion devices, either rack or pedal. Many guitar-type distortion pedals do

Figure 5-9. Leland Sklar.

Leland Sklar on Big Boomin' Bass Tone

"I use Euphonic Audio amps and speakers. Simple rig. Hi-fi. I use the Yamaha sub-kick instead of a mic onstage. It's killer! I'll place it about a half inch off the grill cloth of the single twelve-inch speaker. But I don't use that in the studio. In the studio I use an old TubeWorks DI—normally, *no* effects unless necessary. Only then will I plug anything in. Try to keep things as pure as possible. I bring an amp, but leave it up to the producer and/or engineer to use it or not."

Figure 5-10. Matt Wallace's really old SansAmp units.

Matt Wallace on Taking Low-End Distortion Too Far

"They made two different kinds of these rack SansAmps. On the originals, there was a setting called 'buzz,' and if you turned it up, you could literally watch your woofers extend out from all the low end. There is so much sub-harmonic information. People were blowing things up, and apparently they were calling the owners of SansAmp, telling them that it was ruining their speakers. So what SansAmp did was make a new version that looks the same, but they put filters on the 'buzz' setting so people couldn't blow things up! So if you want the crazy low-end version, you've got to find the really old SansAmps. I have a pair of the originals, because I like the gear that you can actually take way too far."

Figure 5-11. Gnarly bass things.

not work well with low bass frequencies; however, splitting a clean signal in two and putting one of the splits through the guitar pedal, and then blending them both, will give you the edge while maintaining the low end.

USING GUITAR AMPS FOR BASS

You might be surprised to know that the original SVT amps were used not only for bass, but also for guitar! In fact, the first prototype Ampeg SVT amps were sent out with the Rolling Stones on the *Their Satanic Majesties* tour in 1969 in the United States to be used for both guitar and bass, when the band got nervous about using the Fender backline that was planned. Ampeg promised full stacks, plus a full-time technician who would travel with them to swap out tubes for the whole tour. Ampeg custom-made six SVT amps for the tour, far more than were actually needed, with the extras on standby in case there were any problems. The tour was an astounding success and so was the star amplifier of the tour: Ampeg's brand-new SVT design.

With that in mind, never hesitate to plug a guitar into a bass amp (but watch that you don't blow out your speakers from the extra low end). In fact, the Fender "Bassman" head, which has a spectacular history as a guitar amp, was originally designed for bass! Jim Marshall developed his guitar amplifier designs to compete with the popular Bassman and first supplied his amp designs for Pete Townshend and Ritchie Blackmore using circuits originally designed for basses!

Figure 5-12. Bill Wyman with early Ampeg SVTs.

Figure 5-13. Shelley Yakus with Sylvia.

Shelly Yakus on the "Lady Madonna" Bass Secret

"While recording John Lennon's solo album at Record Plant, he let me in on a Beatles recording secret. He said there was no bass on 'Lady Madonna'! It's actually a six-string guitar with the low string being played like it is a bass. Recently, I heard the song on the radio and, sure enough, I could hear the low string on the guitar blended and EQ'd to take the place of a bass."

Figure 5-14. Nevercore's three-string bass.

Figure 5-15. Greg Frederick demonstrates the curious, many-stringed Dingwall.

STRINGING THEM ALONG

Now, I'm bound to piss off a few bass players, but really: Do bass players ever need more than three strings? Or even two? I worked with a clever Japanese noise-core band named Nevercore whose bass player had actually modified his bass to hold only three strings. He had changed the bridge, the pickups, the headstock. I thought it was brilliant.

The Presidents of the United States of America is a trio that has a bass player playing only two strings on a special modified "basitar," and the guitar player playing only three strings on his special "guitbass." I know it sounds kinda stupid, but wow, it totally works.

STRING CHEESE AND HAM

OK, my saying for bass is always "less is more," but in the hands of a creative instrumentalist,

Figure 5-16. Flip-top Ampeg from Linda Perry's collection.

Matt Wallace on Being the King of the Universe

"If I were King of the Universe, no bass guitar in the entire world would have more than four bass strings. But: It would be tuned B, E, A, D. It should have a low B on it, because it's always nice to have the lower notes. But you don't need the G. With the G-string, you are in guitar territory—you should stay the hell away from it. So I would do this tuning so you can go down to a low B if you want to; otherwise, every bass should just have two strings on it. The B and the A string. Truly. Ha ha."

BASS 95

Justin Stanley on His Hollow-Body Kay

"My favorite bass recording setup is a FET (Field Effect Transistor) Neumann U 47 mic on a Ampeg B-15 flip-top amp with my old Kay hollow-body bass guitar. Or to get it a little tougher sound, I'll use an Acoustic 360 folded horn bass amp and cab."

Ross Robinson on the Cure's six-string Bass Plow

"I like having a Fender VI bass guitar on hand. I have one—it's a '62. I use it as a tractor in the middle of a song. Plows right through. The Fender VI is basically the Robert Smith secret weapon from the Cure. I put it through a Vox AC-30, and it sounds amazing, but I need a tech on hand all the time to repair the Vox when it breaks down. Because it's going to break down over and over. I have two guitar cabs that I use on it, mainly—a beat-up-looking one and a brand-new one. One cabinet is a '71 with the original greenback (25-watt) Celestions and Baltic birch wood. The other has vintage 30-watt Celestion speakers."

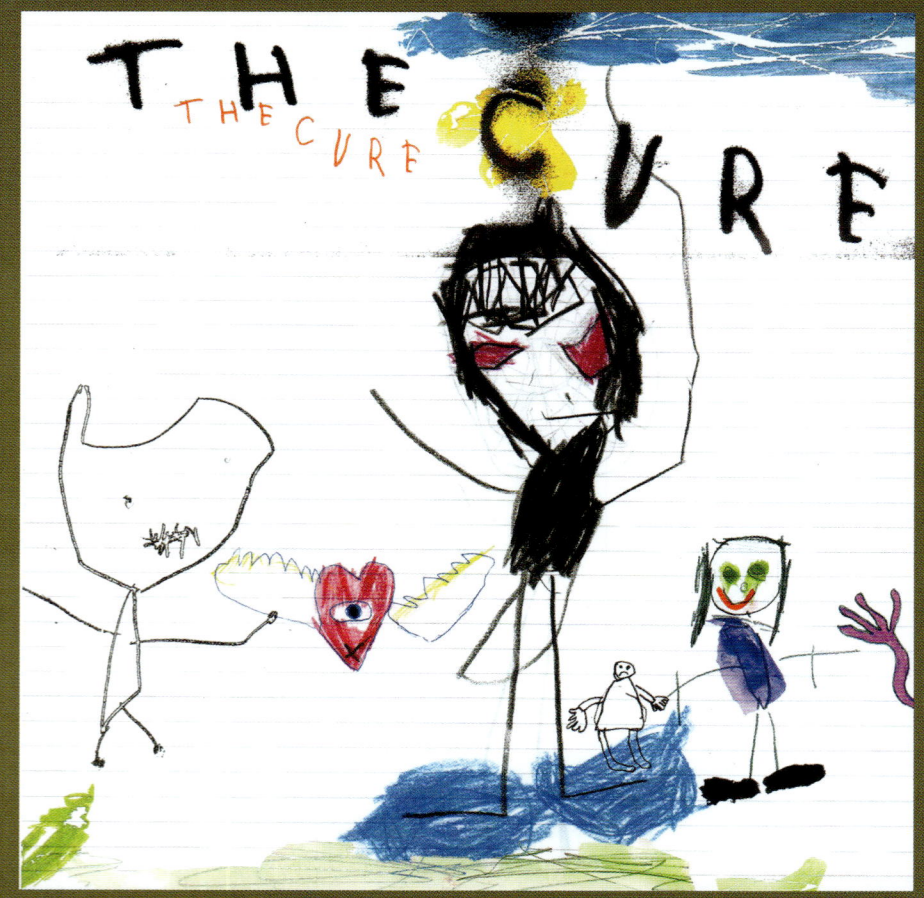

Figure 5-17. *The Cure* album.

Matt Wallace's Butt-Shaker

"We were recording a song for New Beat Fund for Red Bull Records that the bass player was not really keen on. He really had a hard time getting 'amped up' for this song, primarily because they'd already made a couple demos and were, basically, tired of it. So I had him take a break while I set up a subwoofer in the control room, and when he came back, I had him sit on it while he played his part. He'd hit a bass note, and it was shaking in his butt! Because he could feel it, he got excited about the song and—bingo. He nailed the track immediately. Bass players are used to that big thing behind them, right? They like that Ampeg refrigerator speaker (8x10) blasting low end. So I want to design a chair that has a subwoofer built in, so when the bass players sit in it, they could really get off on it! It works for guitar players, too."

Figure 5-18. Here's Flea. He also has a bass.

Figure 5-19. One of these mics should work! Vik Krauss plays the upright.

a super-duper über-multi-stringed bass guitar can be fantastic. Now you just have to find that player!

MORE THAN ELECTRIC BASSES

When you are trying to develop a unique bass sound, another way to go is away from electrified instruments. Acoustic basses can give a track an earthy genuine character, whether they are stand-up or guitar-style. And don't forget about washtub basses!

Figure 6-1.

6
DRUMS

Figure 6-2. Jeff Left of the Girlfriend Experience and his massive Ludwig kit.

WHAT IS THE BIG DEAL?

So what makes drums so important? Consider that if your skyscraper's foundation is crooked and you build on top of it anyway, your walls won't be straight. It is very difficult to fix the foundation once it has been built upon. The same goes for recording music. If your drummer's sounds and performance are not good to start with, piling guitars, bass, vocals, keys, and whatever else on top will not make it better.

Shelly Yakus on Why Drum Sounds Are Important

"If your snare drum sound is just average, it will hold your whole recording back! Many people ask, 'What's the big deal about the drum sound? Why is it so important?' Well, this is what will happen if you don't take time to get good drum sounds. After recording your weak drum tracks, you'll overdub great guitar sounds that will show just how crappy your drums are. Then were you planning on having an amazing vocal sound? Now your drums are going to sound like a midget in there behind all this good stuff. But if you have a really big, important drum sound to begin with—appropriate for the song of course, but where it should be—well, it forces you to make your overdubs better-sounding. Your guitars will be more important-sounding, your vocals will be more important-sounding. It pushes you to get better sounds overall. So the final product going to the mix is a bigger, better-quality recording."

Figure 6-3. Drum setup for the band Hurt.

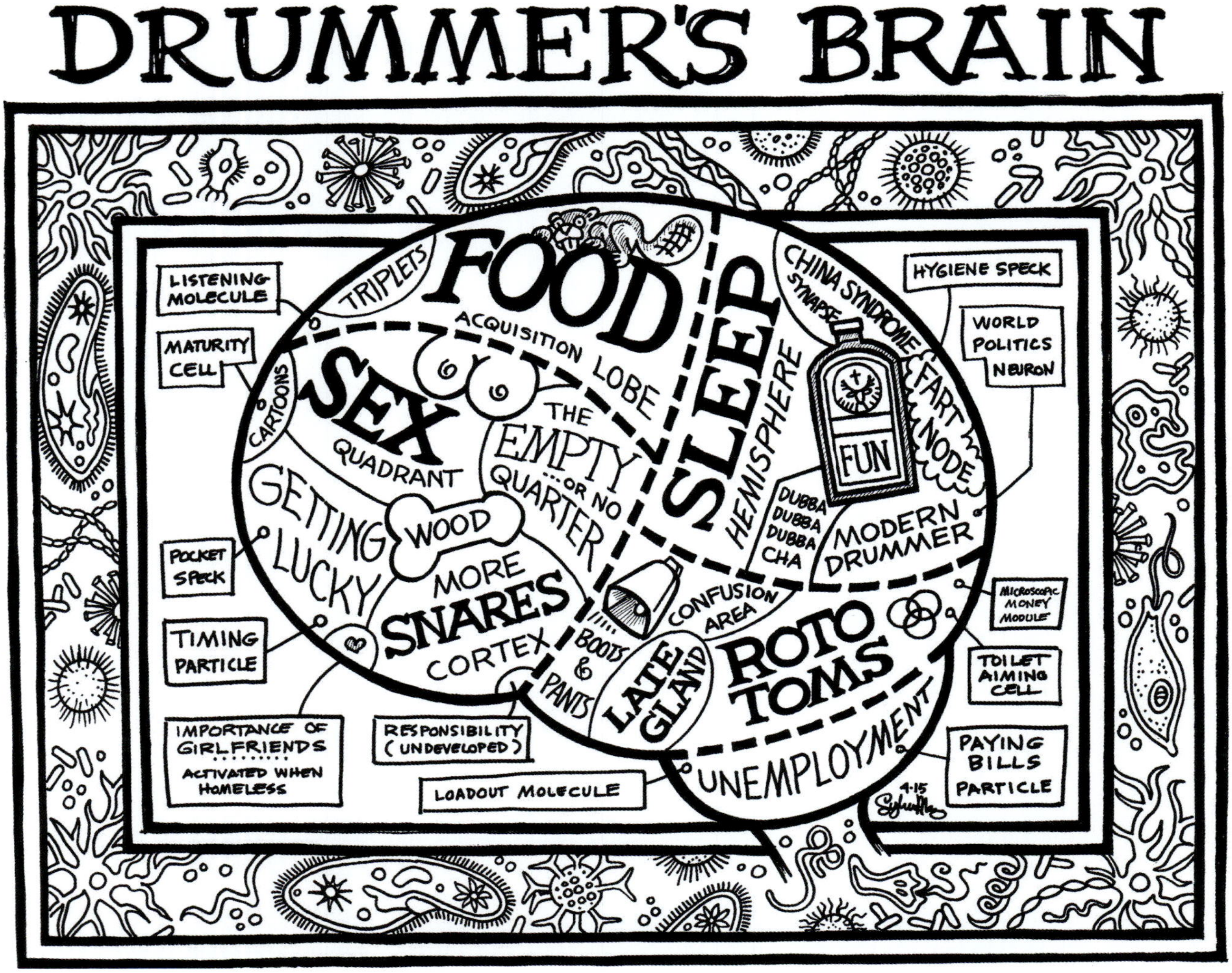

DRUMMER'S BRAIN

Figure 6-4. The Drummer's Brain.

THINK LIKE A DRUMMER

You don't need to play drums to be able to communicate in drum phrases—in fact, most producers communicate in doo-dats, cha cha, bap, doosh, and boops to get their ideas across, and

Susan Rogers on Learning to Groove

"Tommy from Geggy Tah had written a song—it was a love song to my little dog Gina—and he wanted me to play drums on it, because it would be like the mother's heartbeat. And I don't play drums, but I learned to play one groove, and I could basically play it at one tempo! And we sampled Gina's toenails on the linoleum floor, that *ticki-ticki-ticki-tick*, and then looped it. It was like a percussion instrument in the song. It taught me so much about record-making."

Figure 6-5. Morgan Rose of Sevendust has a highly developed drummer's brain.

100 RECORDING UNHINGED

that is OK! But here is a way to really get the trust of a drummer: Know what a drum kit is, how to set it up, and how to play a simple beat. Even if you learn the most basic of boom-chick, boom-chick beats, you will gain a basic understanding of what goes into being a drummer. How the mics affect the way a drummer will play, if they are in the way. How position of the snare and cymbals is a comfort thing, vastly important to a player's style. You need to understand the vacancy required in the drummer's mind in order for them to perform. So I suggest that everyone reading this book take some time to learn how to play a drum kit.

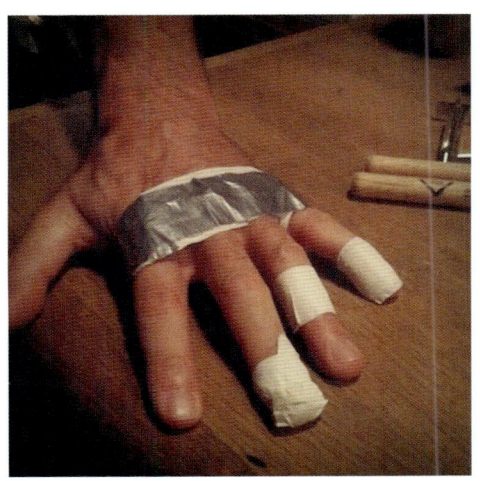

Figure 6-6. Session drummer Dave Watts—wounded but still in the fight.

TRICKS FOR HAPPY DRUM SOUNDS

Hours of the "donk-donk-donk" of getting a drum sound can take the life out of a drummer, leaving him tired and at half energy when it is time to record. If you plan on taking a few hours or longer to get your drum sound, here are some tried-and-true techniques to keep the drummer alive. First, start early and have an assistant to help test mics. I consider it to be abuse of a drummer to have them sit and check mics all day, and then expect them to be excited and energized when the red light is on. An assistant can help you go through the mics one by one, and you can get basic levels for each individual drum by having the assistant or a friend hit the drums one at a time. When all mics have been checked and all phase relationships have been listened to carefully, it is time to bring in the drummer. But it is not necessary to have the drummer go through each drum. I prefer to have the drummer play a simple beat—nothing fancy, just kick, snare, hat for a few minutes while I double-check the mics. Sometimes this is hard for a young drummer to do, because they want to show off a bit; but rein them back and keep it simple. You will want to really hear the air as well as the drum hits, so listening to the space is crucial. Next I'll have the drummer add new elements into the beat: first, cymbal crashes, then tom fills, until I have checked and adjusted the basic EQ on all the mics—this shouldn't take more than ten minutes. Then I'll go ahead

Shelly Yakus on Cutting Holes in Rugs

"On a hard floor you'll find the sound of the tom-toms thin out, especially the floor tom. You get this kind of hollow sound. But when there is a thin rug underneath the drum kit, it tightens everything up, but then the snare drum suffers. So what I used to do is cut a hole in the rug under the snare, so the bottom of the snare drum could see the floor. It would brighten the snare, but keep the rest of the kit tight. After the session we'd turn the rug over, put the patch back in, and tape it up before we got caught! The studio owners didn't like that. So then I would get a bunch of little blocks that I'd make out of two-by-fours—just cut them up with a saw. And I would pile them around the legs under the snare drum stand, and it did the same thing as cutting the hole in the rug. Suddenly that snare would just come to life!"

and let the drummer rip on the kit! Record it and play it back for them, make final level adjustments, and we are ready to record! We may have spent several hours getting a drum sound, but without tiring the drummer out for more than fifteen minutes on the kit.

USE YOUR HEADS

LEARN TO TUNE DRUMHEADS

This is critical: Learn to tune drumheads! Many drummers just change the heads and get all the tension rods about the same tightness. Well, it might sound good in the room, but slap a mic on the top of that drum and you will hear a virtual chorus of notes resonating off of that skin. I have developed several techniques, some quick and some more detailed, to get a great drum tone by tuning.

Here is how I do it: First, replace the head. If you are replacing a snare head, then as you soft-tighten the tension rods, release the snare wires so they are not engaged. As you listen to the head while tuning, you will not want the wires to be tight. Then push your index finger into the middle of the head and, with your other hand, tap the drumhead next to each tension rod. Go around the drum and listen to the differences in tone. By damping the head with your finger in the middle of the drum, the drum will not ring out, masking these tones.

Make a mental note of the pitches across the entire drum, then slowly adjust the tension rods so they all match. When they are tuned, they all sing in unison! Then when you release your finger from damping the top head, the drum will resonate beautifully without confusing cross-tones. Snap the wires back on and "bap"! You've got a beautifully tuned drum!

Figure 6-7. Drum candy.

Ross Garfield (the Drum Doctor) on Tuning Drums

"I tune the bottom heads lower than the top on the toms, and the bottom heads tighter than the top on the snare. In general, that's my technique. I want the drum to have its own character relative to its shell, so I try to let the shell dictate the tuning. You can't make a DW drum sound like a Gretsch drum. And you can't make a Ludwig drum sound like a DW. They're different shells and different animals, and I try to work with that."

TUNING TOMS TO THE KEY OF THE SONG

Whenever there is a passage of a song that has a repetitive tom pattern, the drums sound so much better when they are tuned in the same key of the song. Find the core key of the song and take some time to tweak the toms into that key, where the drums are actually played as a melody instrument. If you are tracking drum tracks for several songs all in one session, it is absolutely worth the extra time to customize the tuning for each song. This was an essential technique on Danny Carey's drums on the first two Tool records.

Figure 6-8. Lucius Borich from Cog is ready for a tuna sandwich.

Figure 6-9. Danny Carey blasting through the *Undertow* sessions.

WANT EXCITING DRUM SOUNDS? CHANGE HEADS OFTEN

There is nothing better than a fresh set of heads in the recording studio. Once a rock band has spent a day cutting tracks, many times the heads will be cratered and dull. You can bring back the snap by changing out the heads as soon as you see wear on the heads. On Tool's *Undertow*, we changed heads between every song—and in most cases, between every take! Danny Carey is such a hard hitter and the songs so long and complex, with so much tom work, it was necessary to keep fresh heads on the kit to maintain the crispness of his drums. On the other hand, if you want a soft, dynamic sound, you may not need or even want fresh heads.

Figure 6-10. Phil Collins.

REMOVING BOTTOM HEADS

Generally, "concert toms" do not have bottom heads. They are a quirk in the drummer's universe. Not really as popular since the 1970s, but still something to try if you are looking for a clean, punchy tom sound of a different flavor. Just remove the rims and tension rods from the bottom of the drums and try miking from underneath—OK, it sounds like shit. Forget I ever wrote about this. Sorry, Phil Collins.

Ross Garfield on George Clooney's Time Machine

"I got a call from George Clooney one time. He was directing a movie called *Leatherheads*. They had tracked the lead song with a major-league session drummer, using his drums. The guy played it perfectly, but George said, 'The drums don't sound right. They sound too new. The movie is taking place in the '30s.' It's simple. I told him I'd bring 1930s drums! And I brought a kit that had the original heads on it from the '30s. Every hit on the drums would bring you back in time. They have a certain sound."

Figure 6-11. "Drum Doctor" Ross Garfield.

Ross Hogarth on Ziggy Marley's Outer-Space Drums

Figure 6-12. Ziggy Marley's *Dragonfly* album.

"*Dragonfly* was a special album for Ziggy. This was his first as a solo artist away from his brother Steven or his sister in the Melody Makers. Here he is, son of Bob Marley, and we were doing a song called 'True to Myself.' As the producer, I envisioned the classic soul sound of Motown or Stax or Volt, but then I also wanted to get an urban groove feel with the emotion of Nas or early hip-hop. I thought the kick drum should sound like an 808, but I didn't want to use a sample or a Roland 808 drum machine. I wanted it played. So drummer Brian MacLeod and I went on this mission to find a kick-drum sound with a long hit like an 808 but still punchy with an attack, and found just the right sound in an 18-inch floor tom! So then the challenge was 'How do we get this floor tom to become a kick drum?' Somehow it had to be played like a kick drum with a kick pedal, while laying on its side. In the studio, we wound up building a cradle for this floor tom; we built it out of bricks and plexiglass. We actually got the thing to work, and it made the song very unusual. Because the kick wound up sounding like it came from outer space, we ended up using a headed tambourine in place of a snare drum. So we created a drum kit on that song that was unique and really different."

DIRTY, DIRTY CYMBALS

Generally speaking, the thinner the cymbal, the brighter it is and the shorter the decay. Cymbals do not need to match, so go ahead and make a cymbal salad. I use a much heavier hat in the studio, because it will be relatively quieter and not bleed as much into the snare mics. A smaller diameter, but heavier hat, like a 13-inch, will have a higher pitch, but not be nearly as obnoxious as a thin 14-inch. To me, dirtier cymbals seem to record better, because the super-shiny clean cymbals will resonate more freely on the high frequencies, which will often overpower a drum recording. Super-clean cymbals are perfect for live shows or for ruining the hearing of all your bandmates, but can be aggravating in the recording studio. Paiste, known for extremely bright cymbals, actually came out with a "dark" series, which they obviously just dumped a bunch of dirt all over them. If you find yourself in a session with a set of way-too-bright cymbals and no alternatives, try taking them outside and rubbing dirt all over them. (Your drummer will just *loooove* that.) In fact, I've created a "dressing" for cymbals that creates a dirty patina when spread over the brass and left overnight. If you're truly fearless, you might experiment with this solution to see how it affects the sound of your cymbal metals.

SYLVIA'S SALTED CYMBAL RECIPE

- 1.5-ounce bag of potato chips
- 1 cup of white wine vinegar
- 1 cup of damp coffee grounds

Directions:

1. Clean your cymbal with steel wool or a pot scrubber to remove varnish if necessary.
2. Crush a 1.5-ounce bag of regular potato chips into a blender. Add vinegar and coffee grounds. Blend until completely mushy.
3. 3. Spread mush all over the top of the clean cymbal. Cover with plastic. Leave in a ventilated area (outside) until sun-cooked, about four hours, then brush the dried chip-spread application off the cymbal.

Figure 6-13. Jim Spiri demonstrates the proper way to defile a cymbal.

If left overnight, the cymbals will achieve a lovely greenish patina. The effect of the spread is to "dirty up" the sound of the cymbal, reducing its sustain and sweetening the harshness of its top end. This effect is reversible, so clean your cymbals with Brasso or a solution of baking powder and water after the session if you desire. Remember, the brown of age on a cymbal can actually be a *good* thing, so if you like the effect, set it with a light dusting of hair spray. Be sure that while you are working on your cymbals, they are set on a towel or other soft surface. Dings and edge dents turn into fractures, and that will really wreck your cymbal. But a little dirt is just fine.

Mark Rubel on Using Mics as Sticks!

"I generally regard recording equipment as musical instruments and don't like to see them abused and destroyed. However, I do sometimes have drummers play with a pair of Nady StarPower mics instead of drumsticks, for the ultimate in close miking. I bought the mics for $9.95 as a 'Stupid Deal of the Day,' but still feel guilty using them that way. Turns out, they're pretty good-sounding mics!"

Figure 6-14. Joey Waronker sez, "Don't come near my cymbals!"

DON'T POINT THAT STICK AT ME

The bigger the stick, the louder the sound—yep, that pretty much sums it up. Hit anything using a chopstick, you'll get a little tap. Use an ax handle, it's gonna make a big noise. I'm amazed at the size of some metal drummers' sticks. Aaron Rossi from Ministry practically uses logs, they are so thick—but you probably won't need logs to get the point across in most recordings. For rock music, 5Bs are generally a good place to start. The thinner 7A drumsticks are better for finesse and detail. They are faster and sweeter. Nylon tips add more definition to each hit of the drumhead and tap of a ride cymbal. So select drumsticks that are appropriate for the performance desired, and have lighter and heavier alternatives available to match for the right feel in a particular song.

WHAT TYPE OF BEATER?

If you are trying to get a drummer with a weak foot to sound like they have more power in their kick, try adding a kick pad onto the kick drumhead, or switch to a plastic beater, or both. Kick pads are a good way to get more attack, but if you've glued a kick pad on the head and want to take it off, then you'll probably have to replace the entire batter head of the kick, as you can't peel off a kick pad without it leaving a sticky residue.

Peter Schickele on P.D.Q. Bach's Fruit Mallet

"On the *Classical Talkity-Talk* album, I used the Pachelbel Canon played on Renaissance instruments. When we got to the session, I felt that it needed a beat–preferably, a lugubrious beat. We were recording in an old theater in New Jersey, and somebody noticed that there was a bass drum sitting on a ledge above the stage. We got it down and it was perfect, except that there was no bass drum beater. But we did find a snare drum stick, and somebody had brought along an apple to eat during break, so we jammed the snare drum stick into the apple and came up with a perfectly fine bass drum beater."

Figure 6-15. Peter Schickele's P.D.Q. Bach.

I like to tape other things to the kick drumhead for a similar effect. Try taping a coin to the head, right where the beater touches. If you are trying to reduce the attack of a kick, try going to a felt beater. Felt beaters will generally sound a bit softer and poofier. If you want it softer, wrap the beater with fabric, or take the beater off and stick an orange on it, or a rubber dog toy! An Indio, California–based company called Drum Concepts makes an assortment of changeable drum beaters made of brushes, "hot-shots" (bundles of small sticks), leather crops, big poofy puffballs. Tons of fun!

PHASE IS GOD

The secret to recording great drums is not really the choice of mics or how many there are—a great drum sound is in the phasing of the microphones with each other. If your drums sound thin and papery, but your intention is for them to sound large and fat, look at how the mics are placed on the kit. If there are two mics facing toward each other without making a phase adjustment, the result will be

Figure 6-16. A phase challenge.

a loss of body in the sound of the drum. This happens most often with the snare, because to close-mic the many elements of a drum kit, the snare mics always seem to be in an awkward position, often out of phase with the rest of the kit; plus, the snare often has a mic on the top and a mic on the bottom, not perfectly facing each other, but slightly off angle.

Listen carefully to your drum mics, monitoring them in mono, two mics at a time. Switch phase on one mic back and forth. Do like an optometrist and ask, "Is it better like this, or like this?" while switching phase. Choose the phase settings with the fullest sound, go to the next two mics, and so on. Change mic position if necessary and check again. If the mics are generally pointing in the same direction, you'll have better phase, no matter how many mics are on the kit.

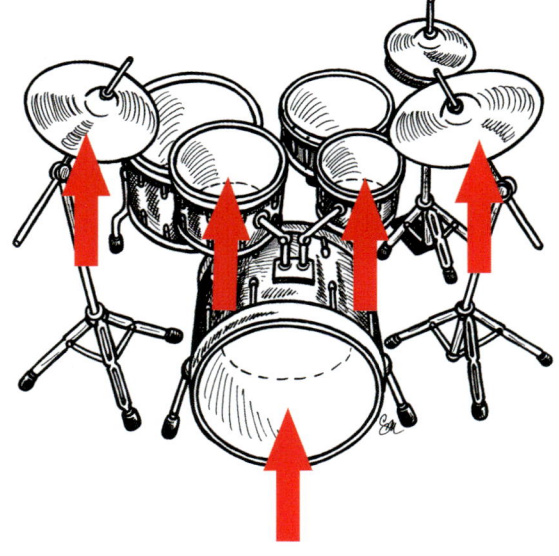

Figure 6-17. General direction of microphones for best phase on drum recordings.

Figure 6-18. Aphex 602 unit.

Paul Wolff on the Aphex and Electro-Voice Drum Recipe

"We were an Aphex dealer and had a blackface Aphex 602 rack-mount processor at our studio in D.C. When I'd record drums I would take two Electro-Voice RE-20 mics for overheads and run them through the Aphex to tape. It had the most enormous low end you could imagine! Normally you might not use the big dynamic RE-20s for recording cymbals and overheads, but this was so good, I wouldn't even mic the toms! It was a bit scary, though, because it was like hanging fifty pounds of mics above the drummer's head. It sounded just like expensive tube condenser mics, because it had all the sweet grittiness on the top end from the Aphex, but the low end was this controlled thunder!"

Ross Hogarth on the "Sherman Filter Bank"

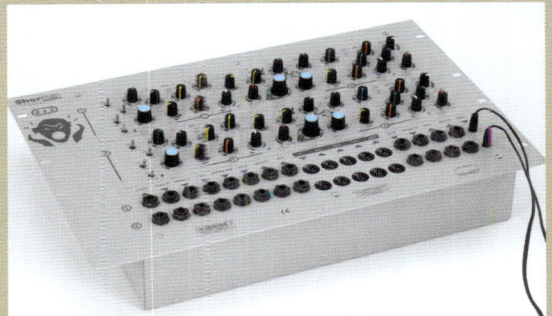

"One of the things I like to use is a 'Sherman Filter Bank' to get drum distortion. I put a bullet mic or some kind of really lo-fi mic behind the drummer and then feed that into a heavy distortion-generating filter. Like a Sherman Filter. The Sherman can really erode the Pacific coastline with its amount of distortion. You actually start to see houses falling off into the sea!"

Figure 6-19. Sherman Filter Bank.

BUILDING A SUB MIC FOR YOUR KICK

One of the coolest devices I've built—and which I use on nearly every drum session—is a "sub mic." It is basically an old retired woofer, pulled out of a speaker enclosure, wired up to work as a microphone! I suspend the woofer by wires, hanging it on a shorty mic stand,

and place the speaker an inch away from the front of the kick drum. I'll sacrifice an instrument cable, cutting off one end and soldering the two leads to the speaker. The other end of the cable goes into a direct box. The sound generated from the "sub mic" is a low, similar to an 808 kick-drum sound. Boomy and extra deep. I'll record this on a separate track from the direct-miked kick drum and blend them together in the mix. A little bit of this sub kick sound gives it extra power and depth. I've also used this "sub mic" on bass guitar for the same enhancement. Yamaha and Moon Mic make similar products to get the extra low stuff that most mics miss.

JACK JOSEPH PUIG ON SUB MIC DANGERS

"The album Jellyfish *Bellybutton* was where I started recording with speakers. I used JBL, Altec, fifteen-inch, twelve-inch, ten-inch speakers. And I started using them like they were microphones on bass drums, bass amps, and guitars. You can capture amazing sounds this way, but remember: These types of 'microphones' are very lethal and dangerous, because the low-frequency spectrum is fairly wide, and you can easily get in trouble. A lot of times, the speakers you are monitoring on can't even reproduce the low frequencies you are collecting. You have to get a handle on high-pass filtering and be very careful."

Figure 6-20. The "sub mic."

ONE MIC, TWO PERSONALITIES

The Audio-Technica ATM250 is one of those genius ideas that makes absolute sense. The single mic housing contains a dynamic with a tight hyper-cardioid pattern, perfectly aligned with a second condenser microphone element that captures the tone of the drum. Put them both together and you get a kick drum with great attack and a big round tone. Fantastic!

Figure 6-21. Audio-Technica's dual-element ATM250DE mic.

SEXY POSITIONS

If you are lacking a good crack or a good body tone in your snare, the solution may be as simple as the position of the snare mic on the top head. By positioning the snare flat across the head, you will get more of the stick sound on the head. By pointing the mic down into the drum, you will get more of the sound of the body of the drum.

Figure 6-22. Snare microphone angles: (a) More attack. (b) More body.

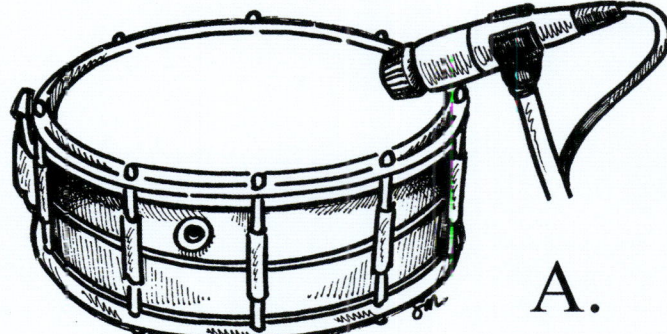

A.   B.

MIKING THE KICK BEATER

A more natural, unique-sounding kick drum can be achieved by putting a mic on the drummer's side of the drum kit, right next to where the beater hits the kick drum head. It will give you an organic "real" sound, but unfortunately, this technique has a few drawbacks. First of all, you will have a lot more bleed from cymbals, hat, and other drums than you would by putting the mic inside the kick drum. Second, you might hear the sound of the drummer moving his foot on the kick pedal. This can be reduced by lubricating the kick pedal so it doesn't rattle or squeak and by having the drummer wear short pants to reduce rustling of fabric.

The third problem is that it can be difficult to get a drum mic into this position and the drummer may have to be conscious of not hitting the mic accidentally, which may restrict his movement in playing. If you can

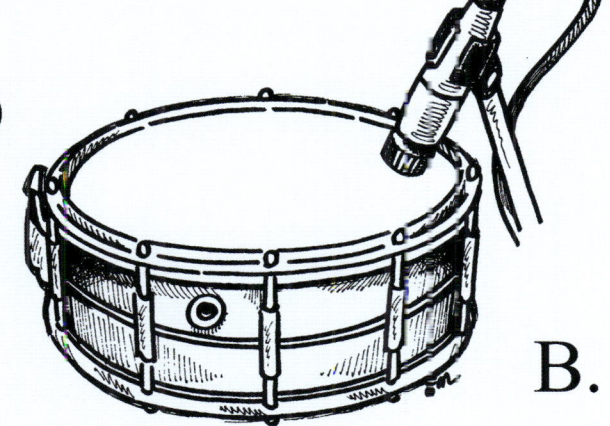

Figure 6-23. Using a tom as a kick drum.

Nick Launay on Kate Bush's Drum Cannons

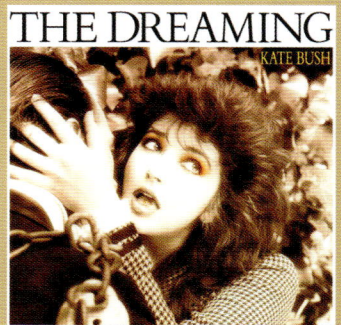

Figure 6-24. Kate Bush's *The Dreaming* album.

"I engineered an album with Kate Bush called *The Dreaming*. It was the first album that she produced. In her wonderfully high voice, having consumed lots of hashish and chocolate during that day, she said to me, 'Hey, Nick, can you make the drums sound like distant cannons?' So, y'know, you try the reverb and it just wasn't big enough. So we had the drums in the stone room at the Townhouse at Studio 2, and then we opened the door to this very long hallway that went out to a loading bay.

"The band the Jam had actually brought in some corrugated iron to put around their drum kit to make it sound like it was in a garage, and it had just been lying around there since their session. So we took the corrugated iron and made this tunnel—a very long tunnel with a succession of microphones, something like four microphones at different distances. Well, it was just a stupid idea, y'know, but it actually sounded good! I mean, we used, like, two pieces with microphone stands and pieces of sticky tape, and you'd hit it, and there would be this kind of reverberant sound. It didn't sound as magnificent as I'd hoped, but it was weird, and it was weird enough to be good. And then I think I even delayed the mics further with some tape delay."

Ross Garfield on Slipknot's Sweaty Drummer

"I was there as the tech for every drum stroke on that Slipknot session for *.5: The Gray Chapter*. Tracking drums took six weeks. Most cats who play double kick don't hit the drums that hard. But this guy Jay Weinberg is hitting the drums and tearing up heads all over the kit. The heads were only good for about three takes—I'd have to change all the skins every hour or so. And he would break at least one cymbal a week. Really good drummer. Every time he'd play a take, it looked like he was losing two pounds in water. He would come out of there after an hour just drenched, and the seat on the throne was soaked through. I'd have to bring in another throne so I could sit there and tweak the drums. Just a pile of sweat. Disgusting."

---

get around these obstacles, miking the kick drum batter side is an excellent technique. I like using it on simple drum kits with wide-open access and few cymbals to bleed. If bleed is welcome as part of the sound you are trying to achieve, definitely try this out. It is especially good for capturing drum performances in which dynamic feel is important.

DAMPING DRUMS WITH SCISSORS AND BICYCLE TUBES

Several drumhead manufacturers make damping rings, but really, they are so simple to make that you need not spend the dough. It is basically a ring cut from a drumhead. When you lay it on top of the drum, it dampens just enough to remove the obnoxious overtones from the shell, creating a crisp snare hit. Turns a clanky metallic "boing" into a tight, snappy "pop."

An outrageously awesome kick drum damping system is an invention of producer Paul Kilmister, son of Motörhead's Lemmy. He stuffs a deflated bicycle tube into the kick, pressing it against the batter head. He connects a tire pump through the front head of the kick drum and pumps up the tire enough for the desired damping effect. Want more damping for a tight double-kick part? Just pump it up a little more! Genius!

Figure 6-25. Paul Kilmister's adjustable kick damper.

Figure 6-26. '70s Ludwig Black Beauty.

THE SURE THINGS

I'm lucky to have some amazing snares. But on most sessions, I'll bring out the trusty Ludwig Black Beauty. Ninety-nine percent of the time it wins. Something about those '20s-to-'70s era brass Ludwigs that rarely miss. Go get 'em.

Ross Garfield on Tom Petty's Star Ludwig

"On the Tom Petty *Wildflowers* album that Jim Scott recorded, they used an old snare, four inches deep, 15-inch head. '20s Ludwig brass drum. It had a lot of character, it has a lot of ring, very old-fashioned-sounding. Steve Ferrone was playing, and the snare was really recognizable. For a while I was on a jag, just buying every vintage Ludwig brass snare I could find, so I have a whole pile of them. When I work with T-Bone Burnett, that's the kind of stuff he'll use, too."

Figure 6-27. Tom Petty.

Ross Garfield on Drum Materials

Figure 6-28. Ross Garfield's magic vintage brass Ludwig.

"It's surprising how good aluminum drums sound, because it is a relatively inexpensive material. A lot of Ludwig drums are actually made out of aluminum, with chrome or other finishes over them. Aluminum will give you a softer, warmer sound, because it is a softer metal. The really strong metals like titanium tend to be on the brighter side. Dunnett makes a titanium drum. The same properties apply to wood drums. Maple seems to be the accepted norm for a drum shell with bright attack, although I have single-ply walnut drums that are also good. The harder the wood, the higher the pitch, and the higher the frequency the drum will resonate."

Figure 6-29. Dunnett's monster snare, made out of an iron pipe.

Ross Garfield on Nirvana's "Terminator"

"One of the most recognizable snares in my collection is the 'Terminator.' It was the snare that was used on Nirvana's *Nevermind* album. It's a bell-brass drum that was actually cast in sand as opposed to being rolled. So it's not only really loud, but it's very musical. You can tune it high and still get some bottom out of it. We used two of those on the Slipknot record. One tuned a little higher than the other. For the faster tracks, we used the snare with the higher tuning."

Figure 6-30. Drum Doctor's "Terminator" Tama snare.

Figure 6-31. Miking the snare port.

MIKING THE SNARE PORT

This technique was taught to me by Mark Needham while I assisted him on a session for jazz great Bobby Hutcherson. Mark would put three mics on the snare drum: one on the top, one underneath, and a third pointed directly at the port in the snare drum. Most snare drums have a port—it is there to allow the pressure to escape when a drummer hits the head. Putting a mic up to the port captures the little "puff" attack that comes out of the drum, making a crisper hit when blended with the other mics. It is a pretty cool technique, but trying to keep hat bleed out of a third snare mic caused me to abandon this as a regular practice. It is a good thing to try, especially in quieter recordings, to really capture the subtleties of a snare performance.

GOING SHOTGUN

Shotgun mics are useful in live settings when you have to record from a distance, but they are

Shelly Yakus on Hanging Mics from Tall Ladders

"I got up onto the highest ladder I've ever seen. It scared me to death. It was A-shaped, and then had another ladder that went straight up the middle and kept going. It felt like it was a thirty-foot ladder, but not leaning against the wall. Producer Roy Cicala had me go up to the top and hang a shotgun mic off the ladder over the drums to get the snare. Yes, the drummer was under this huge ladder. I was terrified going up there, but it actually sounded really good."

also delicious in the studio. They can be positioned above the snare, and then blended into the kit to give the snare a halo of space without interference from other drums and cymbals.

NOTHING IS PERFECT

Most rooms are not perfect—or should I say, *no* room is perfect. You make do with what you have. I might want to record in a large hall for a big, exaggerated drum sound and a small room for a tight, intimate drum kit. I might need both these sounds for the same song, so I look for recording spaces that are as versatile as possible.

Ross Hogarth on Mötley Crüe's Drum PA Technique

"We would go down to the Complex Studios, which was owned by George Massenburg. He had the PA systems that were used for Little Feat and Earth, Wind & Fire, set up on a big production soundstage for rehearsals. That area was completely separate from the recording studio, which had a dead drum room. So we'd record the drums in the soundstage instead, setting up the PA and sending the kick, snare, and some of the toms back into the room to get this massive explosion of sound. The toms get hit, and all of a sudden, the room is being filled up with the PA. It was a massive sound. It didn't take a whole lot, but it made a huge difference.

"We used this technique on Mötley Crüe's *Girls, Girls, Girls* album on Tommy Lee's drum sound also. It was a different studio—One on One on Lankershim in North Hollywood—but we did the same thing with the PA. It's a drum sound that you can only get by recording it in a bigger room. You can't get this sound by conjuring it up afterwards with fake reverb and stuff."

Figure 6-32. Mötley Crüe's *Girls, Girls, Girls* album.

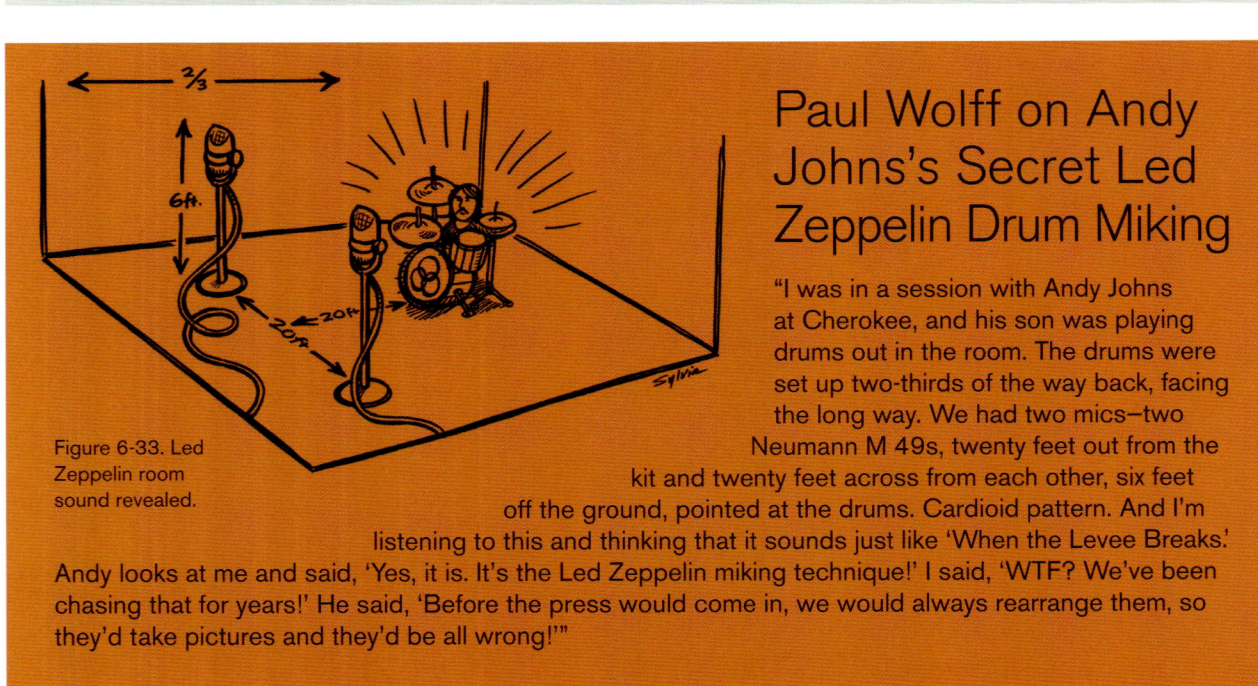

Figure 6-33. Led Zeppelin room sound revealed.

Paul Wolff on Andy Johns's Secret Led Zeppelin Drum Miking

"I was in a session with Andy Johns at Cherokee, and his son was playing drums out in the room. The drums were set up two-thirds of the way back, facing the long way. We had two mics—two Neumann M 49s, twenty feet out from the kit and twenty feet across from each other, six feet off the ground, pointed at the drums. Cardioid pattern. And I'm listening to this and thinking that it sounds just like 'When the Levee Breaks.' Andy looks at me and said, 'Yes, it is. It's the Led Zeppelin miking technique!' I said, 'WTF? We've been chasing that for years!' He said, 'Before the press would come in, we would always rearrange them, so they'd take pictures and they'd be all wrong!'"

Brad Wood on Small Rooms

"For drums, most of the time, I really like small rooms and parallel walls, because not many people I know hang out in perfectly designed rooms with bass trapping that goes underground and all those bizarre angles. When the drums are in a big neutral room, I'll be compressing the crap out of them to get some excitement, and I'll get nothing. Same as putting the drums in an open field. It's just boring. Now it might be different for strings or other instruments where you don't want a lot of engagement with the room. There you want a big neutral space. I want a listener to hear a loud rock band and feel that there is something really urgent and they need to pay attention."

Figure 6-34. The joy of recording in a small room

Ross Garfield on Hans Zimmer's Twelve Drummers

"On the score for [the 2013 Superman movie] *Man of Steel*, Hans Zimmer wanted to hear twelve drummers playing at once. He described this concept to me and just let me set it up the way I thought best. . . . So each drummer wound up with a drum set, two timpanis, and an orchestral bass drum. And Drum Doctors provided almost all of the rentals for the session. The drummers were Jim Keltner, Josh Freese, John Robinson, Jason Bonham, Matt Chamberlin, Pharrell Williams, Danny Carey, and more. Literally the world's best drummers. We recorded it at the big Sony scoring stage in Culver City. It took two days to load everything in, set it up, and get drum sounds. And then the drummers showed up. Hans gave the drummers a cadence and they'd start. 'Bamp, bamp, ba-ba-bamp da-da bamp, bamp, ba-ba-bamp da-da . . .' The director standing in the middle of the room would give the drummers cues for dynamics, bringing them down soft and bringing them up as hard as they could play it, cued off the screen, playing that cadence. As they brought the intensity from soft to loud, you could just imagine an army coming over the hill! Hans didn't want everyone to be 'Pro Tools perfect.' He wanted the slop in there, because that's what made it sound real. Three drum kits were the most I've ever done at one time before this session. But to have twelve guys playing at the same time was crazy! When they all played it loud, I would get extremely nervous, because at that point they're denting and breaking the heads, and then it was my job to run around and retune every drum set!"

Figure 6-35. Hans Zimmer's twelve drummers for the film score for *Man of Steel*.

Paul Wolff on Creating an Easy PZM Effect

"You can take a Shure SM57 and put it right up against a wall and it will work the same as using a PZM mic. The PZM is a 'pressure zone mic'—everyone thinks it is real scientific, but the whole object of a PZM is that there are no reflections; there's just the direct impact of the sound into the mic. If you face the SM57 right up against the wall, maybe an eighth to a quarter inch away, as close as you can, the mic will only pick up the energy that hits that wall. Same thing. If you set a mic up in the middle of a room with a drummer playing, it will pick up the reflections from all the walls, making it sound very confusing. If you focus that mic directly against one wall, you get none of those confusing reflections."

My last studio was in a cavernous theater. To control the drum sound, I built a truss system that I could hang canvas panels from to create different-size rooms. I'd often set a drum kit up in the lobby of the theater to get that tight, intimate club sound. Sometimes I'd set a drum kit up onstage and pull the curtains, keeping the cymbal wash out of the big room.

Another great way to control cymbals is to record them separately as an overdub! It takes a moment for the drummer to get used to playing on towels draped over the cymbals and hi-hat, but wow—later on when you crank up the room mics and there are no obnoxious cymbals, you will love it!

WEIRD DRUMS

So there are some pretty strange drums out there. Drums that fold up so you can carry

Figure 6-36. Jim Keltner plays a Trixon Speedfire kit.

Figure 6-37. Trixon Speedfire drum kit.

Figure 6-38. Trixon Cocktail drum kit.

them to gigs. Melted drums that look like they sat out in the sun too long. Drums that are built like tuba bells. Drums made from plastic, glass, wine barrels. Some of my favorites are Trixons, which use forward-thinking technology in their designs. The Trixon Speedfire was first introduced in the '50s and mystified musicians with its misshapen kick drum, which actually accommodated two separate chambers played with two separate kick pedals—making the Speedfire the first double-kick!

Trixon today offers a "cocktail kit" that is perfect for small stages and is played standing up!

Spaun makes fabulous acrylics, reminiscent of the classic Ludwig Vistalite drums of the '70s. My Spaun acrylic snare is so loud, it cuts through the most bombastic drummer's performance.

Someday I hope to write a book all about weird drums—North drums, Dalbess drums, Staccato drums, Aural drums, drums kits made of timpanis, drum kits made from tambourines. Drum kits made of rounds of cheese. Someday I will. Someday!

Figure 6-39. Spaun clear acrylic snare.

Figure 6-40. A snare made out of a tambourine!

PREPARED DRUMS AND ALTERNATES

If you want really unusual drums, open them up and stuff them with rocks, pennies, nuts and bolts, paperclips, unpopped popcorn, candy wrappers, used guitar strings, car keys, broken drumsticks, etc. Or tape items to the rim and let them bounce around. That rattle-y stuff will add to the excitement of the drum kit. Just don't put in the drum key—you'll need that.

Figure 6-41. Prepared drums.

Was it just a rumor that Neil Peart had beer bottles lined up during sessions, which he grabbed and hurled into a miked-up box, breaking them in time? Well, if it isn't true, then it should be! Adding more than just drums into a kit will make for some interesting moments in the studio. On more than one occasion I've added a Jaymar toy piano into a drum kit. Bud Gaugh from Sublime played a melody on the Jaymar during his performances on the SexRat record. Pure fun!

Mark Rubel on Alternate Drum Kit Ideas

"I regularly use a postal shipping box for a bass drum, pizza box with change and chains for a snare, pie tins for cymbals. Sometimes I'll do the 'no-drum kick drum': the beater of the pedal hits a Shure Beta 52 directly! Sounds pretty good with EQ. I also built an electronic drum trigger into a rubber chicken, which was hilarious."

Figure 6-42. Sublime's Bud Gaugh plays the Jaymar in his kit.

DRUMS 121

Figure 7-1.

7
GUITAR

Figure 7-2. Whitty from Spiderbait's lovely SG.

ELECTRIC ATTITUDE

Deliciously messy guitar rigs make me hot. What do I think about super clean perfection? Boring. Give me two or three or more amps all daisy-chained together, a big paper bag of pedals dumped out on the floor, cables and wires attached with alligator clips.

Ed Stasium on the Secret to the Ramones' Sound

"The secret to the Ramones' guitar sound? One-hundred-watt Marshall, one cabinet, everything on ten. Just turned up, with all the knobs actually gaffered that way. And Johnny Ramone's Mosrite guitar, straight in. That was it. That was the setup. And of course there was Johnny. That was the real secret to the Ramones' sound. You can buy John Bonham's kit and set it up exactly the same as Bonham. You could play that kit, and it ain't gonna sound the same as Bonham playing it—no way. You can get a Marshall and turn it up to ten, and it just ain't gonna sound like Johnny Ramone—same thing."

Eric Valentine on Tape-Machine Guitar Color

Figure 7-3. Greta's Scott Carneghi phones home.

"I like tape machine distortion on guitars. I'll run a guitar signal in, just the direct signal, right into the mic input of a tape machine to get all the coloration, which is much more of a Motown-ish kind of sound. While the guitar player is performing the part, they don't get to hear what the tape machine is doing, because there is a delay to it. Afterwards, we line up the recording with the rest of the tracks for a listen. The tape machine is a really great source for distortion on clean guitars. I'll do this technique with songs people send me to mix, too. I'll process the guitars through the tape machine and then lay it back in time in the session."

CURIOUS WOOD AND WIRE

There are some peculiar guitars out there to experience! Some of my favorites were designed by an innovator named Vincent Bell. He created the Coral sitar, which delivered a unique electrified sound in the '60s. He also built the Bellzouki and Longhorn basses that defied convention in the early days of rock 'n' roll.

Other notable offbeat stringed critters are the baritone guitars, like the Danelectro bari and the Schecter five-string. By using them to reinforce the rhythm guitars in your recordings, you'll raise the height of your sonic spectrum. I use the baris doubled, spread wide in a mix, playing just the root note of the chords I am reinforcing. Not much distortion—just a brace to hold up sometimes too-frizzled guitars.

Figure 7-4. Vincent Bell's Coral sitar.

Figure 7-5. Schecter five-string baritone guitar.

Ed Stasium on the Dan Armstrong Guitar

"When I was a kid, I saw Keith Richards playing that plexiglass Dan Armstrong guitar. I said, 'I gotta have me one of those!' I forgot exactly how much it cost, but I do remember I had to trade in my '63 Gibson SG Custom, plus pay another $65 for that Dan Armstrong guitar. Well, I got rid of the Dan Armstrong soon enough. They're horrible guitars. Interchangeable pickups and they all sound the same. They're heavy. They're really thin-sounding. Terrible guitar. I always regret making that trade."

Figure 7-6. Dan Armstrong guitar.

AMPLIFIERS UNHINGED

One of the strangest amps I've found was the "Tuned Tube" by Acoustic. It sat neglected in the attic of the Chicago Store in Tucson, Arizona, covered with pigeon droppings. The handsome old thing has two reflective panels that are adjustable on either side with knotted ropes. This was one of Heartbreaker Mike Campbell's favorite amps on the Johnny Cash *Unchained* sessions.

Figure 7-7. The Acoustic "Tuned Tube" amplifier.

Figure 7-8. Making noise with the Teisco tabletop amp, speaker pointing straight up!

I've had some wicked-good luck digging through back rooms of ancient music stores, discovering old forgotten amplifiers. I dust them off, change out tubes, and replace—whatever it takes so they are ready to terrorize the studio!

THE NEW WEIRDOS

There are some bizarre and wonderful new amp/speaker designs to keep things interesting. A fellow named James Scott makes three-sided speakers to go with his 3rd Power amplifier line. Do they sound good? Who cares! They look *cooool*!

So good they might be illegal! Tone Tubby is an outfit out of Marin County, California, that makes speaker cones out of hemp. No prescription required!

Figure 7-9. 3rd Power's unique speaker design.

Figure 7-10. Tone Tubby speaker cab, barely legal.

Ross Hogarth on Van Halen's Variac

"It's funny for me, having made the last Van Halen record, that Variac was very much what Ed (Van Halen) was all about, and he still is all about the Variac. So, if you starve the voltage, if you starve the tubes, you'll get that certain tone. The whole trick to the killer sound is actually all about starving the amp for power. I've never ever fried anything by messing with a Variac. You have a certain amount of play before the transformer is going to collapse and just not work. I have had shit just shut down because it is not getting enough power to operate, and I've had fuses blow when I try running it hot, but nothing gets hurt. And more often, you are going for less voltage 'cause that's where you really get the tone. So you can try a whole range of Variac settings, but the sweet spot might be somewhere between 105 and 110 volts, at least ten voltage units down."

Figure 7-11. Variac units.

TUNING A HAM-FISTED PLAYER

Typically, when certain chords in a guitarist's rhythm performance are out of tune, but the rest of the performance is in tune, the instrument's intonation is instantly suspect. But wait a minute—maybe the blame should be put on the guitar player instead! Here is how you can tell. Tune several guitars carefully using an electronic tuner. Have the guitar player play the same part on the three different guitars. If the same chords are out of tune on all three guitars, then you are dealing with a "ham-fisted" guitar player, which is a guitar player who learned to fret improperly. They may be pressing too hard on the strings, pushing the strings to one side instead of pressing directly down, or just pulling the strings in ways that make certain chords a problem while other chords are fine. There is seldom enough studio time to retrain the guitar player to fret properly. But here is a solution to get the recording done: Start by tracking the part, warts and all, and identify the problem chords. Then go back and retune the guitar while the guitar player is fretting the problem chords, and punch those chords into the performance. That's right—get the guitar player to fret the chord, and while they are holding the fretted position, have them play each string one at a time. Then you look at the tuner and you adjust each string so it is in tune. Then record that chord everywhere in the performance it appears. Cut and paste digitally if that works, too. It is a tedious process, but there is nothing more powerful than multilayered rhythm guitar parts, tight and in tune.

Figure 7-12. Cecil Gregory hams it up.

Larry Crane on Stringing It Nashville-Style

"A 'Nashville guitar' is a regular acoustic guitar strung with all the high strings. It's a standard tuning, but it sounds much brighter and more colorful. The beginning of Tom Petty 'Free Fallin'' has that sound. It feels Byrds-y, with a Roger McGuinn vibe to it. You can use this type of guitar to tuck into mixes, and it lifts the whole track. I will record two passes of a regular six-string, then layer two passes of the Nashville on top of it, split wide, and it's got a real jangly thing like a twelve-string but different. Bigger. I had a cheap Yamaha acoustic guitar, and I dedicated it to be the 'Nashville guitar' at our studio. I got a luthier to rebuild the nut and the bridge so the thinner strings would sit real well. Got it set up properly, then put a P-Touch label on it that says, 'Do not change strings!'"

Figure 7-13. Slide National Triolian.

Matt Wallace on Relearning to Play Guitar

"There's this band Howlin' Maggie who were on Columbia Records. They had done a demo of this one song three times, so when they finally got around to doing a guitar solo for the real album, [guitarist] Andy Harrison was like, 'Wow, I've already done this song three times.' He was just playing 'guitar by numbers.' So I just took his guitar and put it in a really odd tuning. It wasn't out of tune, but it was in an unusual open tuning. He started playing it and became really mad at me.

"Well, he played and struggled and played. It started really sketchy. He had to relearn how to play the guitar, and he eventually came up with this performance that, in the end, was a really inspired solo! He actually couldn't physically play what he had written for the solo before on all the demos. He couldn't even play guitar the way he usually plays. The guitar was not his friend anymore, so he had to go in entirely different directions."

Figure 7-14. Producer Jim Wood adjusts Scotto Landes's sound while recording guitar for Mankind Is Obsolete.

TINY AMPS, BIG SOUND

Bigger is not always better. Spiderbait's platinum album *Tonight Alright* was entirely done with this tiny "micro-twin" Fender amp, so small it was dwarfed by the Sennheiser 421 microphone used to record it. I have a collection of a dozen little guitar amps that comfortably take on big roles in almost every album. Some of my favorite lines include Zinky's "Smokey" amps, Pignose, Valco, Silvertone, Piggy, and Airline.

Figure 7-15. Micro amps make lot of noise!

Matt Wallace on his RadioShack Amp

"You'd be amazed at how loud this tiny Realistic amp is. It's just a little RadioShack miniature amp and speaker, barely even one watt, but it will drive a 4x12 Marshall cabinet no problem. It's remarkably loud!"

Figure 7-16. Matt Wallace's noisy little micro thing.

Ed Stasium on the Ramones with Phil Spector

"Johnny Ramone would not work with Phil Spector unless I came along to the sessions, because I had been involved with all the Ramones' records up to that point. They trusted me. So even though I was not producing on *End of the Century*—because Phil Spector was—I came along as the band's 'musical director' to the sessions at Gold Star Studios. I would normally do a lot of 'ghost' playing on the Ramones records, and this was the same on *End of the Century*. Johnny would play his Mosrite through his Marshalls, and I used my Strat and played through Johnny's little Mike Matthews Freedom amp turned all the way up. The Freedom amp was a battery-powered thing you could take on camping trips if you wanted to. Made by the guy who designed all the Electro-Harmonix guitar pedals. Phil had us playing electrics together at the same time, recorded to separate tracks. Then he had us put acoustic guitars on every song, sometimes twelve-string. He would record us together on the acoustics, always to a mono track! I always wanted a Phil Spector *Back to Mono* button. Jeez, I'd been a fan of his sound even from before I knew what a producer was! So I kept asking him for a button, and on the last day of the session he gave me one. I really liked him."

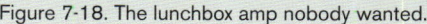

Figure 7-17. Ed Stasium's gift from Phil Spector.

Matt Wallace on His "Barbie" Amp

"This is an amazing little guitar amp in a lunchbox—I bought it at a flea market down at Fairfax in L.A., and it sounds wicked! And there's still enough room to put your lunch inside it. This guy in Los Angeles made these things. It's all hand-built! He had some much, much cooler ones, but they were more expensive. This was the cheapest one he had because no one wants a Barbie and the Beast lunchbox!"

Figure 7-18. The lunchbox amp nobody wanted.

Figure 7-19. An AEA ribbon and a vintage Electro-Voice 666 with some sweet combos.

Figure 7-20. 12-Gauge Microphone.

A PANTRY FULL OF MICROPHONES

Microphones are the conduit between what you hear and what goes on your recording. As with cooking, once you have the ingredients (guitars, amps, speakers), there are many ways to prepare a meal. The most daring chefs have a variety of cooking tools at the ready, and they are not afraid to detour from any recipe. So my mic closet has the tried-and-true standards and a whole bunch of oddball mics, too.

Some of the coolest mics I've discovered recently are 12-Gauge Microphones—built out of repurposed shotgun shells! Collections of four shells come in a little ammo box, ready to record some killer sounds!

Ed Stasium on His Muddy Mic

"I had a Shure 555 that I bought when I was a kid. It eventually fell out of favor and was stored in my mom's basement for years and years. Unfortunately that basement was seriously flooded in the early '70s, and my old 555 was underwater for three days! Well, I got the mic back and shook it out; it eventually dried out, and I just kept it as a memento. A piece that I put in my cabinet. It was still all full of dried mud. I never really cleaned it out properly. It wasn't until 1999 that I actually plugged it in to see if it still worked, and, sure enough, it did. Not only did it work, but it actually sounded really good! So today, that mic is my go-to guitar amp mic—seriously! There's no top end on it. No bottom end on it. It's just pure middle, without any of that top-end frazzle that you get with a condenser mic or even a SM57 or a Sennheiser 421. It honks! It's actually 'muddy'-sounding!"

Figure 7-21. Ed Stasium's muddy mic.

Figure 7-22. Yeah, good luck with that.

PHASE FLIP AND COMMIT!

If you have more than one mic on a guitar cabinet, find a blend that works for the part you are recording, have all the mics sum to the same track, and commit them to a blend. Be sure to try to reverse the phase between the mics to find the blend that is the richest, fullest. I will flip the phase back and forth by using a phase reverse mic cable, or if your mic preamp has a phase switch on it, do the phase check there. Switching phase and blending mics in the box never sounds nearly as good as blending and committing before recording to the digital

Larry Crane on Summing Mics

"If you put two mics on a guitar cabinet and record them to two tracks, it never sounds quite right. If you put those two mics and blend them onto one track it always sounds great! How can that be? Well, generally Pro Tools is going to always be a little out of phase. I'm always having these issues, so it's much easier and better to sum the mics while recording."

recorder. By blending mics, checking phase, and moving mics when needed, it's easy to record a great guitar sound without ever engaging an EQ.

MICROPHONE ROBOTS

Eric Valentine has a particularly clever way of finding the sweet spot on a speaker cone, while having the ability to align phase among several mics on the same cabinet. Yes, that is a remote-control robot device. And yes, he built it.

Figure 7-23. Eric Valentine's guitar robot.

THE SPLIT-AMP TECHNIQUE

On guitar day in the studio, it's time to strategize how the guitar tracks will be built on any particular song. I find that setting up a split-amp arrangement allows the most versatile guitar recording. Basically, such an arrangement consists of the guitar going into a splitter box and then out to two separate guitar amps, which drive two separate guitar speaker cabinets. I choose two guitar amps that have different but complementary characteristics—say, a Rivera Knucklehead Tre and a Bogner Ecstasy, with the Tre set to be super compressed and the Bogner set to be very dynamic-sounding, like an old Marshall 100-watt lead. Next I'll mic the cabinets and bring the mics to blend and, depending on the part being recorded, if it needs to be more aggressive, I'll push up the Rivera. And if it needs to be more expressive, up comes the Bogner. I sum the mics together, after checking phase, and record all the blend from one performance to one track. One track at a time!

Figure 7-24. System Of A Down's split-amp setup.

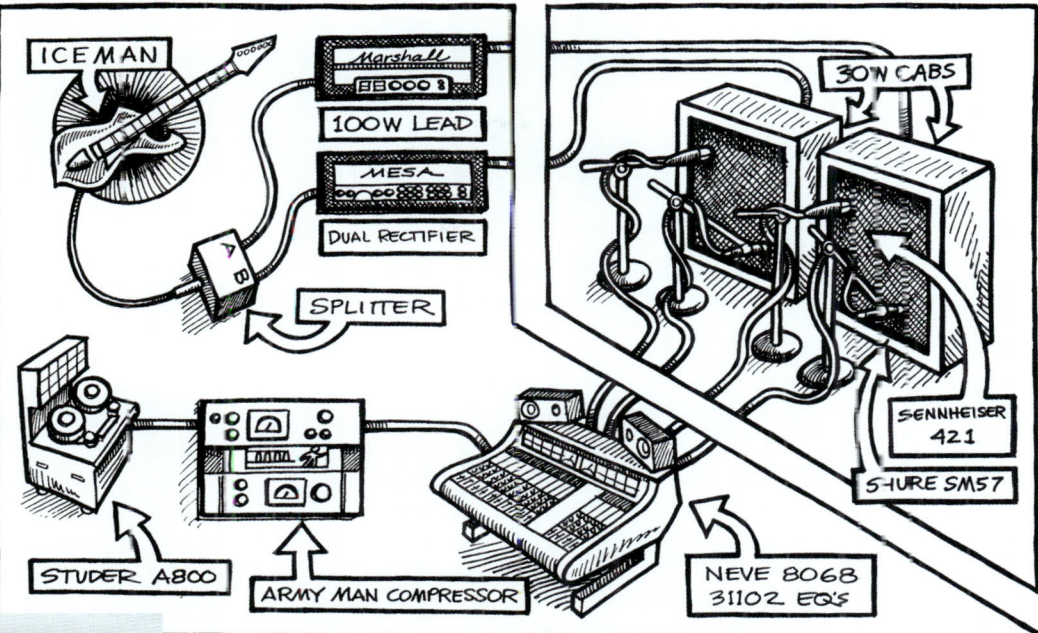

Figure 7-25. Adam Jones from Tool.

For stereo guitar, as we did on Tool's *Undertow* album (the song "4°"), we split the two amps to a pair of tracks—a Marshall 100 watt lead / Mesa Dual Rectifier split. One is a naturally compressed-sounding amp, one a naturally dynamic-sounding amp. Gives you this crazy panning effect that swims around in your head. And there ya go. So do you know what that Tool song is really about? Something to do with the temperature difference between the anal cavity and the vaginal cavity. Yes really.

Ross Robinson on Aengus at Indigo Ranch

"It was strange for me to leave Indigo Ranch Studios and work anywhere else, because that Aengus console there was so clear and transparent and dimensional. You can hear an example on those hand-mixed first Korn records and the first Slipknot record. You could virtually see all the way in the back of the room. The Aengus console at Indigo was 100 percent essential in developing that Korn sound. Since Indigo Ranch, I've never been able to get that again, nowhere. When Richard Kaplan from Indigo sold the console, he ripped the Aengus EQs out and had them sold separately. Holy shit. Well, that was the heart of the whole thing."

Figure 7-26. Legendary Aengus thumbwheel graphic EQ.

GUITAR 133

INDECENT GUITAR EQ

I love me some Neve, but when it comes to guitar, the 1073s might just need a little boost. I add a rack-mount API 550, Orphan Audio Electrodyne, or Aengus graphic EQs. That gives me the control to really get what I want in a guitar recording.

Figure 7-27. Orphan Audio's Electrodyne EQ.

MOVING TARGETS

Leslie speakers give the Hammond B-3 a unique sound, but why use rotating speakers only on organs? With a special pedal called a Leslie Combo Preamp, you can put anything through a rotating Leslie speaker: a guitar, vocal, piano, synth, or whatever. For guitar, it is especially conversational. This is a technique I used on Tool's "Sober" for the main rhythm guitar. Of course, you'll need not only the Leslie Combo Preamp but also a Leslie—preferably a 122 cabinet. Leslie also made a series of rotating cabinets specifically for guitar, currently marketed as the Leslie G37.

Of course, you can also create your own Leslie effect by suspending a guitar cabinet from a rope and spinning it. Or have

Figure 7-28. Do-it-yourself rotary speaker.

Tim Palmer on Finding the "Sweet Spot" in the Room

"During the '80s, the recording process was well funded, and this gave us the chance to perfect our vision. These were decadent studio days—we had lots of time and great studios to work from. We would often spend a good amount of time searching out the 'sweet spots' for mic placement when looking for that special sound. In retrospect, we may have spent too much time, but compared to today's rushed, budget 'lite' sessions, it was at least more creative and fun.

I was recording an album in London with the Mighty Lemon Drops and I often brought my remote-control car to the studio to play around with in the live room. One day I figured that if I attached a mic to the car roof, I could drive the car around the studio, being controlled by myself in the control room, and look for the perfect distance/phase and sound for the room mic. Worked pretty great, as I remember."

Figure 7-29. Tim Palmer's radio-control microphone device.

Mark Christian on Recording the Monkees' Guitars

"I played guitar on a Monkees record—yes, a Monkees record. The producer, Roger Bechirian, went to RadioShack and bought twenty Realistic PZM mics and had them taped all over the big room at Cherokee Studios in Los Angeles. We got on a ladder and taped them all over the walls and ceiling with duct tape. As we were recording the guitars, the tape was peeling off and the mics were dropping, but we just left them that way! It was a loose yet calculated way of recording, and a lot of fun!"

Figure 7-30. Colorful hand-painted ZVEX guitar pedals.

someone run around holding a mic in front of your blasting guitar cabinets. Or put a guitar cabinet on a skateboard and push it around the room. You get the idea!

EVIL LITTLE BOXES

Guitar pedals. Some of the best are built by nutty lone-wolf "Edisons" in their garage labs. One of those guys is Zachary Vex of ZVEX Effects. His masterpieces include the over-the-top "Fuzz Factory," the "Seek-Wah," and more recently, the "Probe." He has them individually hand-painted, which gives them a personal character, and everybody wants them. And they do things no other pedals will do. They are absolutely, disgustingly mad.

Ed Stasium on Using a Hair Dryer to Get More Fuzz

"I used to rehearse in an unheated basement when I was a kid, with my Strat and my Maestro Fuzz pedal. I used to huddle next to a space heater to stay warm, and I noticed that the guitar pedal sounded totally different when it was set next to the heater! It was wider and had more *raarrrr* instead of being so 'zingy.' The sound was so much better that I used to bring the space heater to gigs and set it up onstage so I could heat up the pedal. Years later, I tried it again in sessions by heating up the Maestro with a hair dryer and, sure enough, it really does something special. Seriously, you want to try bringing a hair dryer to your sessions."

Matt Wallace on Building Pedals from Kits

"There was a company called PAiA that would offer kits that you could build, and I built a flanger pedal from one of their kits. Then I built a 'bi-filter follower' from another company. It was basically a volume-controlled filter, so the harder you'd hit it, the farther it would open up, like an automatic wah-wah. I put it together, and the enclosure had wires hanging out of it, and it really wasn't cleaned up. But I used it anyway, and it was great!"

Figure 7-31. PAiA's spring reverb kit.

BE THE GUITAR PLAYER'S FOOT

Never be shy about helping a guitar player out—or, if you are playing guitar, don't be afraid to grab another person to help you out! The most exciting solos are live events, choreographed and performed by more than one person. As an engineer, I've played the wah-wah with my hand and adjusted settings on the ring modulator pedal in the middle of a guitar player's solo, and it has created amazing results!

Figure 7-32. Matt Hill takes over foot duties from Fred Pool during guitar tracking.

Geoff Emerick on Controlling Robin Trower's Guitar Effect

"Robin Trower's guitar was a really hard guitar to work with. We had three Marshall stacks in [Studio 1] at AIR Studios in London. I had U 47s or U 67s on them, about three of them up in the air. Anyway, his sound, with that Fender Blender pedal—which is that sort of 'Hendrix'-style *woooh-woooh*—was uncontrollable. If you watch it on the VU meter—and I always use VU meters because they show you everything—well, the level was going up and down so extreme that nothing would hold it. Finally, the only compressor that held it was that square Neve compressor, the 2254, which is a great compressor. All the other compressors it went straight through."

Eric Valentine on Slash's Octave Fuzz

"When I was working with Slash on his solo record, he was in the process of developing this octave fuzz guitar pedal for MXR. During our recording sessions, there were versions coming in from Dunlop that he was checking and making tweaks to them. We all got to listen to them and comment. It finally got finished, and it's an amazing pedal! It generates one of the most monstrous, aggressive, incredible guitar sounds. There's a song on the record that I just finished where there's no bass at all—it's just drums and guitar with his MXR SF01 Octave Fuzz Pedal on it. And the low end is just massive."

Figure 7-33. Slash.

Figure 7-34. Marius Roth from Seigmen playing a giant slab of wood.

Wah As an EQ

An interesting substitute for EQ is to have an instrument plugged into a standard wah-wah pedal, then set the pedal halfway and leave it in that position. This creates an extreme filtered, nasally sound depending on the position. It's excellent on guitar, of course, but also great on synths and other keyboards, and even vocals. Try re-amping a drum room track through a wah pedal this way!

MORE THAN GUITAR

And of course, these boxes work on more than guitar. Try them on drums, vocals, synthesizer, acoustic guitar, piano—oh heck, why not banjo, tambourine, or marimba! I used a whole series of pedals on a theremin with Fishbone's Angelo Moore. It completely changed my idea of the theremin as an instrument.

Susan Rogers on Prince's Pedal-Driven Drum Sound

Figure 7-35. Prince's *Sign 'O' the Times* album.

"Prince had that LinnDrum (LM-2) machine, and he ran it through Roland Boss pedals. Instead of the individual kick, snare, or percussion tracks, he would take a submix out of all the drums. And then he'd run it through his flanger and his overdrive pedals. Or his distortion pedals and his delay pedals. That was pretty innovative—a lot of people imitated him for that. I'm thinking in particular of the song 'Hot Thing,' from the *Sign 'O' the Times* album. The effect is very, very deep there."

TASTY TALKBOX

Building a talkbox is a super easy, fun thing you can do with a little guitar combo, a funnel, and some plastic tubing. Start by duct taping the funnel over the combo's speaker. Connect five feet of tubing over the end of the funnel, and put the other end in your mouth. Plug your guitar into the amp and play. The sound will travel through the tube up to your mouth. Now, if you have the end of the tube in your mouth and "mouth words" as you play the guitar, you have a fantastic recordable sound. You can also get a similar effect by playing a music file on your iPhone while sticking the little speaker end in your mouth.

Figure 7-36. This is the actual talkbox Peter Frampton used on *Frampton Comes Alive!*

CONTROL ROOM FEEDBACK GENERATOR

Here is a great trick for getting guitar feedback while still having the guitar player standing

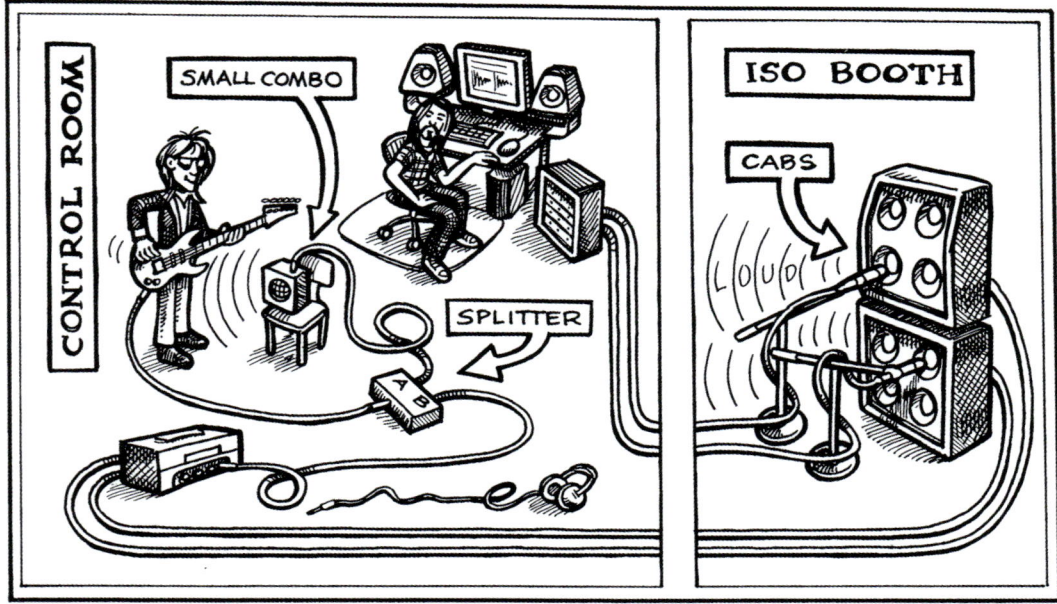

Figure 7-37. Control room feedback technique.

138 RECORDING UNHINGED

Figure 7-38. Matt Davis from Showbread has got good guitar licks.

Figure 7-39. Surface transducer.

in the control room—and no, you are not using the studio monitors to get the feedback, because most of the time it just doesn't sound right. This is what you do: Put the guitar into a splitter and have one lead go to the main guitar rig, which should be out in an isolated room, miked up and ready to go. Take the other lead of the split signal and put it into a little combo amp that you can set in the control room, close to where the guitar player is performing.

Have the guitar player use the little combo to generate the guitar feedback while recording the results from the big rig. It works beautifully and keeps the guitar player from having to perform in the room with a giant guitar rig blasting. But of course, the guitar player might just need all that noise to get himself off—if you know what I mean.

MATT WALLACE ON BUILDING WEIRD SHIT

"When I started in my parents' garage making records, I had to be really creative, because I didn't have money for gear. That's why I made reverb plates out of transducers and stick-on pickups. I thought, *I can't afford to buy a plate, so I'll build one!* Actually, I built four of them. The kind of driver I used on these plates was a 'surface transducer.' It's a thing you can screw onto a wall, and it makes the whole wall a speaker!

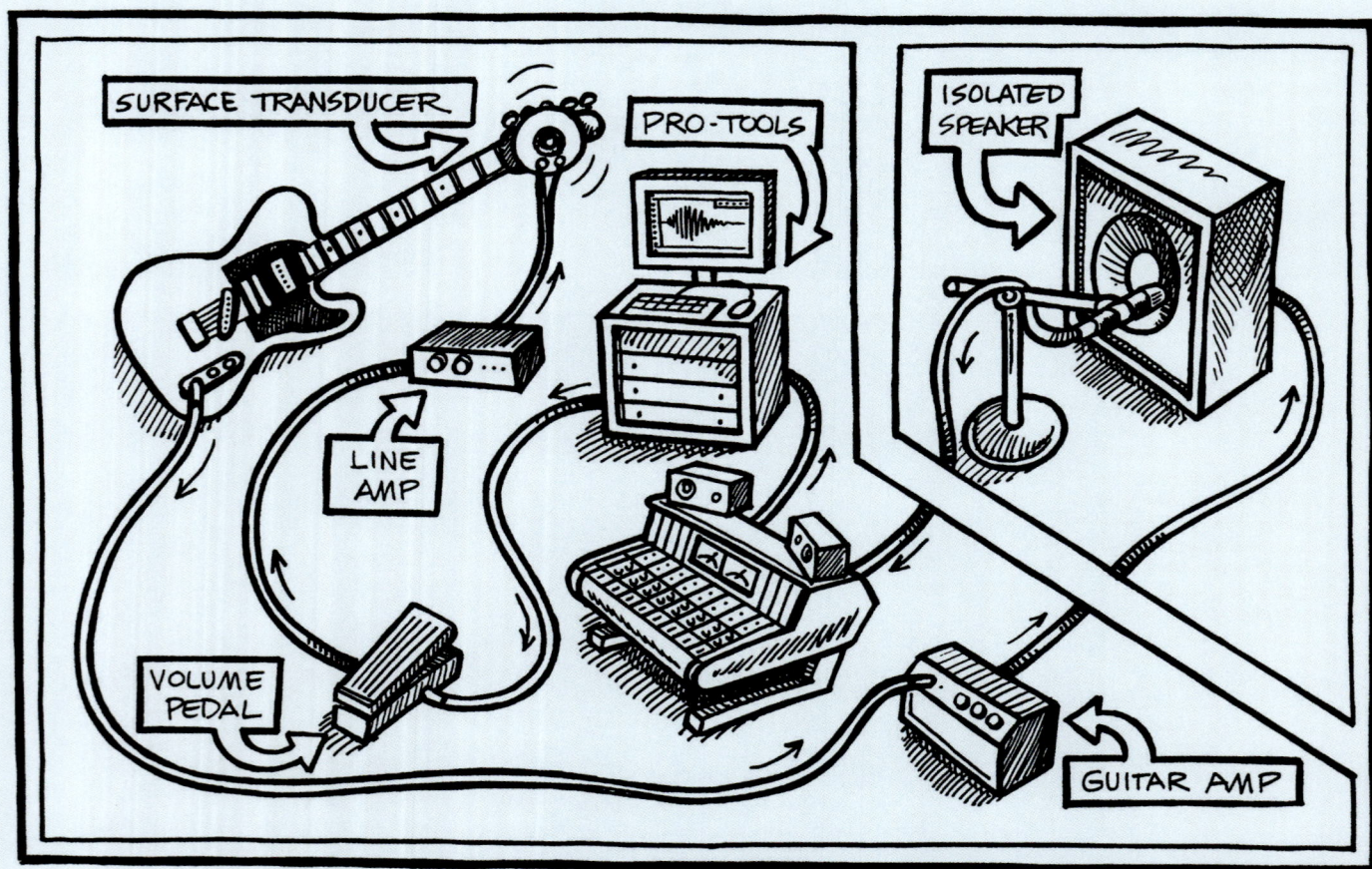

Figure 7-40. Matt Wallace's guitar sustainer.

"So I would epoxy the surface transducer to the plate, and then I had these two Dean Markley pickups that I had initially purchased for an acoustic guitar. I would stick those on the plate with some putty, hoping it would pick up the sound. It worked, but it really didn't sound that good. But here's another idea that I figured out later: I took one of these surface transducers and duct taped it to the headstock of a guitar, ran an amplifier to it, and used a volume pedal as a controller. And you know what that did? If you feed the sound of the guitar output from the console back into this transducer, it makes this crazy sustaining feedback that you control with the volume pedal!"

JUSTIN STANLEY ON GETTING SWEET DISTORTION ON A DIRECT GUITAR

"For the ultimate guitar sound, put your 1963 Gibson SG through a direct box into a Neve 1073 mic pre input, gain up the mic pre just before it pinches, then put that into a Urei or Universal Audio 1176, then out of that into *another* 1176, both of them full on . . . then patch that into another Neve 1073 line in, to get your level to the track of the recorder (analog tape preferred!)."

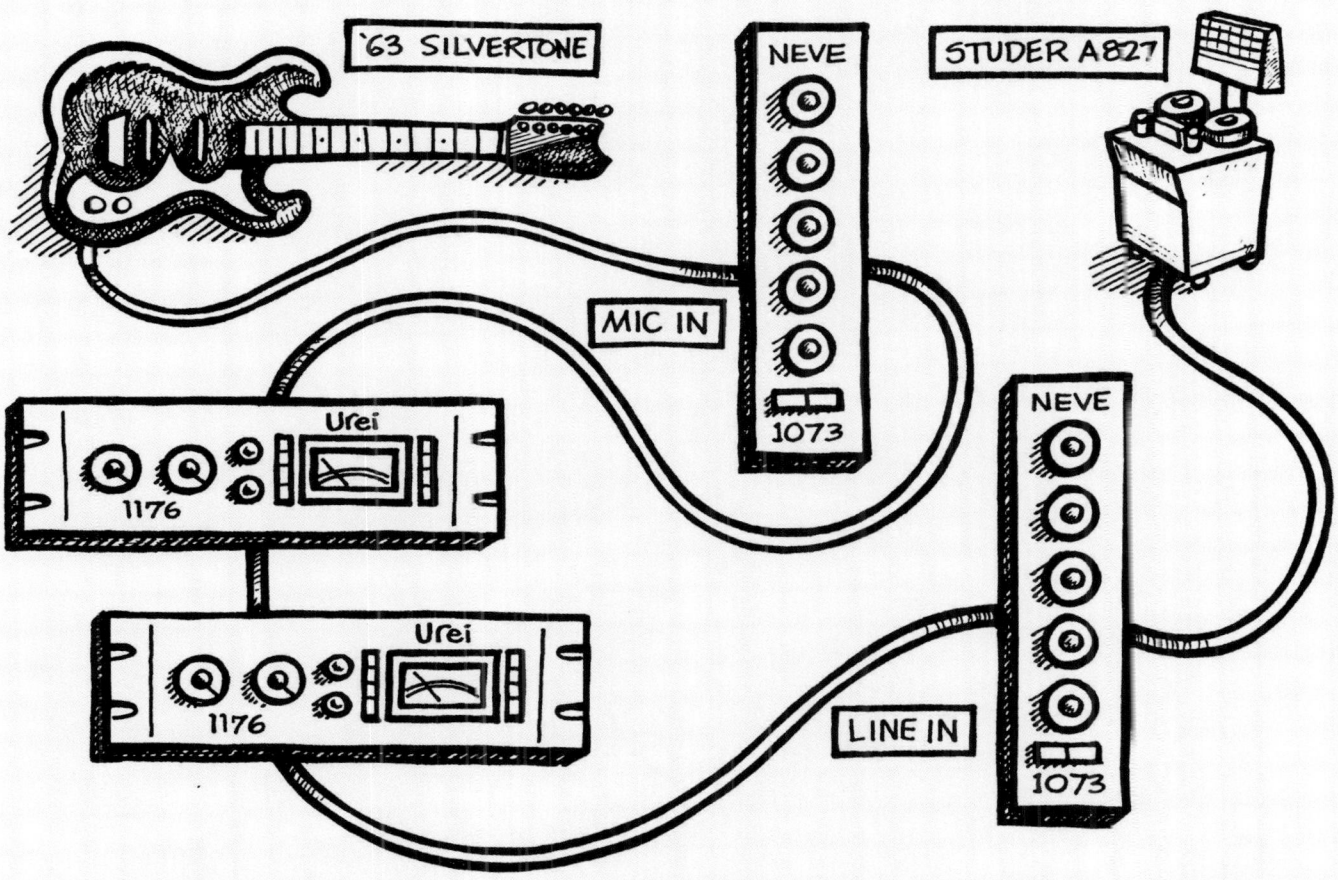

Figure 7-41. Justin Stanley's no-amp guitar diagram.

Figure 8-1.

8
PIANO AND ORGAN

Figure 8-2. Roger Manning.

UPRIGHT PIANO AS A 19TH-CENTURY ENTERTAINMENT CENTER

Before there were radios, television, laptops, smartphones, or recorded music at all, entertainment in many homes centered around the family piano. Just about everyone had one or wanted one, much as is the case with flat-screens and computer tablets today. You could keep up with popular trends by buying the latest sheet music at the corner store and trying it out on your family's upright. Almost every young kid got piano lessons. This is why there are so many old uprights floating around the classifieds today. Many of them you can pick up for a hundred bucks, or even free if you'll cart it off. Even more than a century later, these relics can be used to enhance your recordings, or even star in one of your studio productions. It must be noted that many old pianos are considered "totaled" once their hammers harden, strings go tubby, and tuners get soft. It would cost far more than the piano is worth to repair and tune these pianos, but as experimental sonic canvases, they are amazing! Every "unhinged" studio should have one, or two or three!

THE GRANDER, THE BETTER?

If the core of a song you are recording is the piano, a true grand may be a necessity. Of grand piano sizes, there are mainly two types: the concert grand, which is around nine feet long; and the studio grand, which is around seven feet long. Anything under seven feet is in the baby grand category and will have a shallower, lighter sound, closer to an upright. Of the true grand pianos, the smaller seven-foot piano may be the better recording instrument for popular music. It has slightly more percussive, mid-range quality, and has a more focused sound than the larger nine-foot grands. However, if you are recording Chopin and are super serious about it, better make sure you are packin' the big nine-foot guns.

Figure 8-3. Ben Folds has spent a few grands.

Figure 8-4. Susan Rogers.

Susan Rogers on Vari-Speed Piano

"I worked with a band called Geggy Tah on a song called 'Such a Beautiful Night.' We wanted a piano that sounded wider than a normal piano. So we took the tape machine, which was a Stevens, and used the vari-speed to slow it down while Greg Kurstin played the right-hand piano part. Then we used the vari-speed to speed it up, and on a separate track he played the left-hand piano part. So when you play it back at the fixed speed, it sounds like a wider piano than normal. The high side of the piano is higher and thinner, and the low part of the piano is lower, so it sounds like his arms are extra long!"

Figure 8-5. Miking piano holes with Blue Hummingbirds.

Figure 8-6. Realistic PZM mic under the lid.

Figure 8-7. Miking the back of an upright.

PECULIAR MIKING

Really, why must it be so complicated? Be sure to weigh the importance of technique over inspiration. "You need a large-diaphragm *blah blah* suspended six inches over the hammers on the low and high *blah blah blah*." Come on! Just stick a Shure SM57 on there right above the middle and see what happens. Hey! It sounds damn good! And you are ready to record immediately. Sometimes you just don't want to fuss when the ideas are flying around the control room. Don't lose momentum by trying to get a pristine piano with a perfect stereo image, just so you can bury it behind a wall of guitars. Ha! But on the other hand, taking time to experiment can be very rewarding. While testing some Blue Hummingbird mics, I found that miking the holes in the harp of a grand actually is a "thing" that will give you a round, resonant tone. Not bad!

LID REFLECTIONS

For isolation at Sound City Studios, Elton John had a box built to house the top of the studio's grand piano. By replacing the lid, he could play live with a band in the same room while still isolating the piano mics. Pretty cool apparatus. A piano lid's position can also be used to create and direct the sound out of the piano body. A few well-placed mics using lid reflections make for some nice piano sounds. A piano lid opened halfway and PZM mics taped under the lid are excellent for intimate performances.

Another curious technique is to mic underneath a grand piano, capturing the woody resonance of the soundboard. You can adjust the sound brighter or duller by moving or removing rugs from underneath the piano, too. No rug equals a brighter tone!

GRANDIOSE UPRIGHT

A good, quality upright can actually be a terrific instrument for recording, so when you find one, you drag it to your cave. The piano tuner arrives and seems to take delight in immediately wounding your pride by pointing out the infinite flaws in what you thought was a good deal. Relax, everything is OK—it's what they do. Give them an hour and they might come full circle. True!

An upright will never sound exactly like a seven-foot Steinway, but it may sound better than a lower-end baby grand if miked up properly. Here is how I mic my Mason and Hamlin: I pull the piano away from the wall about eighteen inches and mic the back of the soundboard halfway down. I may use two mics to get a stereo landscape, if the song calls for it. I generally don't open the top and mic the inside, because I find the mechanics to be too noisy and noticeable in the recordings. However, if a song needs a little clanky character, I'll pull the front panel and expose the strings and hammers. I'll set the mic right up close to them and capture the hammers hitting the strings and the sounds of the pedal and action movements.

Justin Stanley on Taming the Eric Clapton Upright

"On the Eric Clapton sessions for a record called *Clapton*, we used an upright piano. Some of the keys would stick on it, so you'd have to use the side of your fingers sometimes to pull the key up again to keep playing—that was with a fellow named Walt Richmond, who was J. J. Cale's piano player, with Willie Weeks, Jim Keltner, Steve Winwood, and the horns from the Dap-Kings. I used to love giving great players those kinds of instruments to play, because they have to fight the instrument—they'd have to tame the beast a little, y'know? It lends itself to a bit more emotion sometimes."

Figure 8-8. Eric Clapton's *Clapton* album.

PREPARED PIANO AND OTHER EXTENDED PIANO TECHNIQUES

Composer John Cage introduced the "prepared piano" technique in his avant-garde *Sonatas and Interludes* collection in the 1940s. Generally, a prepared piano is a grand piano with various objects inserted between the strings—paper, forks and spoons, bamboo, keys, nuts and bolts, and various things. This technique adds percussive effects to piano performances, random buzzes, mutings, rumblings, and taps—very interesting stuff. It is also reversible, as you can take the objects out of the strings and restore the piano's original condition. So, a "properly" prepared piano is one that can be "unprepared." The Velvet Underground and Nico used paper clips on a prepared piano for their song "All Tomorrow's Parties," produced by artist Andy Warhol.

Figure 8-9. Andy Warhol.

PLAYING PIANO WITH SOMETHING OTHER THAN HAMMERS AND KEYS

Other "extended piano" techniques include directly plucking and scraping the strings with fingers or other objects (this is a technique called "string piano"); hitting the outer rim of the piano; yelling or singing into the piano harp while pressing down on the sustain pedal; palm muting piano strings while playing the piano; or tapping the overtone positions on the strings while playing to create harmonics. Composer Karlheinz Stockhausen would have the pianist wear a certain type of gloves while performing his compositions, to protect the pianist's hands while *POUNDING* on the keys. I suppose that also made a difference in the performance. With gloved hands, a pianist would most certainly pound on the piano much harder!

TACK PIANO

A tack piano literally has metal thumbtacks pressed directly into the felt of its hammers, which causes a distinct percussive sound when they strike the piano strings. It is a fun

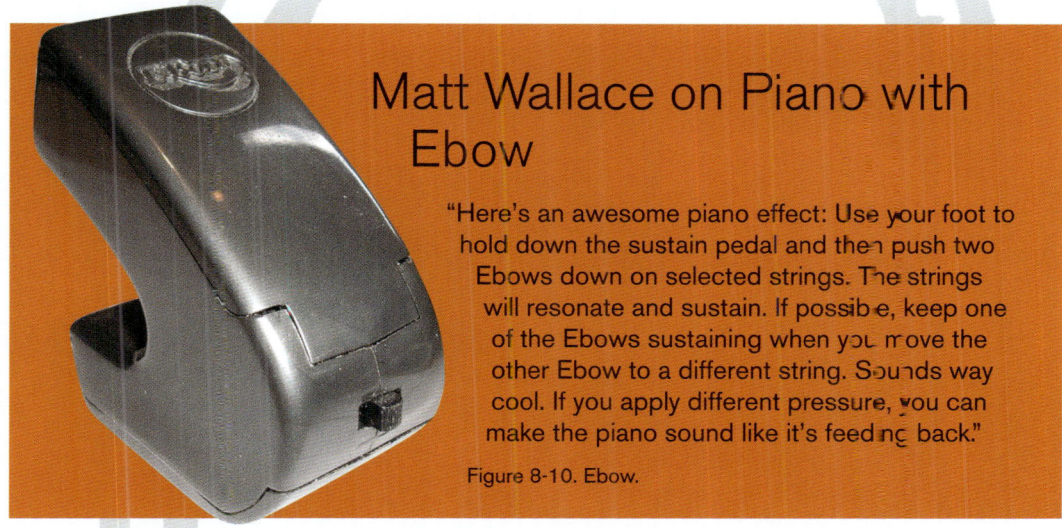

Matt Wallace on Piano with Ebow

"Here's an awesome piano effect: Use your foot to hold down the sustain pedal and then push two Ebows down on selected strings. The strings will resonate and sustain. If possible, keep one of the Ebows sustaining when you move the other Ebow to a different string. Sounds way cool. If you apply different pressure, you can make the piano sound like it's feeding back."

Figure 8-10. Ebow.

"saloon"-style piano sound, originally done to make the instrument (usually an upright piano) loud enough to be heard over the din of a noisy pub crowd. It is sometimes mistaken for a type of "prepared piano," but this technique will damage the piano's hammers and is generally not reversible, so the technique is not recommended for valuable studio pianos.

Figure 8-11. Rick Rubin.

RICK RUBIN'S SINGLE-NOTE IMPACT PIANO

As a way to add impact to any new section to a song, Rick Rubin will add a single low-root piano note on the downbeat of any new section. Sometimes we would double them and split them wide for dynamic excitement. These piano parts are then mixed low in the final mix, sometimes just felt and not heard. Makes the listener feel like something important has happened without being obvious. We used this technique on several projects, including System Of A Down's debut and Johnny Cash's *Unchained* albums. Of course, if you really want to shake things up, record this type of single-note impact piano and crank it up in the mix! Or flip the files backwards—the same thing works great reversed!

CLAV, RHODES, HARPSICHORD, AND WURLI

Besides acoustic piano, there are several electrified piano-like instruments with unique character to add to your recordings, using mechanical weighted keys, strings, or tines and hammers similar to those of their acoustic counterparts. Samples are available for many of these instruments, but it is much more adventurous to drag a real one into your session. The Fender Rhodes has a vibey, phasey piano sound, instantly recognizable. Same goes for the Wurlitzer electric piano. You can occasionally find student models in thrift stores for a decent price.

Figure 8-12. Wurlitzer electric piano, from Linda Perry's collection.

Julian Colbeck on Hauling Around a Rhodes

"My Fender Rhodes was so damn heavy, it was ridiculous. In fact, I used to put the Fender Rhodes in my Mini. By myself. And the only way I could do it was to take the front seat out, actually unbolting the passenger seat. For years I had to do this when I played recording sessions in London in the '70s."

Figure 8-13. Another thrift store find: my massive Hammond C-3.

I recently found a '70s vintage Baldwin harpsichord in a thrift store for cheap and am determined to restore it. Harpsichords are similar to pianos, but the strings are actually plucked with tiny, soft picks instead of hit with hammers. Hohner Clavinets are also distinctive instruments, finicky as hell—but wow! What a great way to give your tracks a unique personality.

HAMMOND CHEESE

People drop them off at my studio like I'm running a home for old wayward organs. They think to themselves, "Oh, this is a good place for my grandma's treasured

Figure 8-14. Benmont Tench, one of Tom Petty's Heartbreakers.

Lowrey." I have Baldwins, Kawais, Conns, Rodgerses, and Hammonds. Little do they realize I'm abusing these instruments—stripping out their innards and reusing their reverb tanks, rhythm machines, and tone generators for other things. Then I'm painting them, hanging them from hooks, and lighting them on fire. Poor things! [*Evil laughter . . .*]

The Hammond B-3 is a whole 'nother animal. I don't hurt that one. It is the dual manual organ with a selection of drawbars that drives a separate two-speed Leslie cabinet. This is the organ sound of Deep Purple, the Allman Brothers, Genesis, and Pink Floyd. Booker T. & the MG's used it on "Green Onions." Procol Harum used it on "A Whiter Shade of Pale." Jon Lord modified his Hammond so it could be played through a stack of Marshalls, which gave it a monstrous, overdriven sound. Bob Marley and the Wailers used the Ham-

Figure 8-15. Colorful Hammond sound.

George Drakoulias on the Muscle Shoals Helicopter

"We were at Muscle Shoals studio before they shut down, with soul artist Dan Penn, and we were trying to use the studio's Leslie speaker, but it was broken. It was really, really bad and intermittent, and they said, 'Oh, let's not use it,' and I said, 'Nah, let me try it.' And I stuck my hand inside to get it to work—it sounded like a helicopter! It was the Hammond organ on a song called 'Zero Willpower.' They said, 'We better fix that,' and I said, 'No, we're going to use it!'"

mond B-3 on "No Woman, No Cry," and it is the signature "bubbles" sound on a zillion reggae songs. The Hammond is a vital ingredient on modern pop recordings, too—you'll hear its chirpy whistles in tracks by everyone from Tom Petty to the Beastie Boys.

I found mine in a thrift store in Yreka, California. It was retired from a local church and was in pristine condition. Mine is actually a C-3, which is a B-3 with a "church cabinet." Same guts, different look. *Waaaay* heavier.

SHOW ME YOUR PIPES

In an age when recording can be done with a pair of mics and a laptop, it might be time for you to rediscover these majestic instruments hidden in churches, halls, and theaters. They need to be appreciated! The phrase "pulling out all the stops" comes from playing these massive machines, because of the way the sounds are controlled with pull-out levers to change tone. Pipe organs use actual pipes made of wood and metal with reeds, operated with multiple banks of keys, air pumps and bladders and bellows. You can imagine the maintenance on these things, and because of this, so many sit idle with walls of pipe banks for show only. But there are several still in operation, and they are a thrill to experience. They use entire buildings as part of the instrument. The biggest pipe organ is in Atlantic City, with 33,114 individual pipes, each with a different timbre and tone.

Figure 8-16. Daniel Roth plays the seven-manual pipe organ at Atlantic City's Boardwalk Hall Auditorium, the largest operating pipe organ in the world.

Figure 8-17. The pipe organ at the Temple Church in London.

Hans Zimmer on *Interstellar*'s Pipe Organ

"The pipe organ is really a huge, complicated synthesizer, if you think about it. You have a pipe and air blows through it, and that makes a sound of one pitch. If you want to add color to it, you add the sound of another pipe, and it becomes these really complex harmonic structures. During the recording of the *Interstellar* film's soundtrack, there was this endless discovery going on for myself and director Chris Nolan. Temple Church organist Roger Sayer played the church's incredible pipe organ, and he would take us through all these different voices and all these different colors."

Figure 9-1.

9

STRINGS, HORNS, AND ORCHESTRA

KEEPING IT OLD-SCHOOL

There is nothing like the real thing—I mean real *string*, har har. Organic strings instead of the synthesized or sampled kind. Life, breath, and astonishing spontaneity manifest themselves when you work with live humans. Human emotion channeled through a violin is euphoria for this rock-'n'-roll girl. The sound of many string voices together in a "choir" is otherworldly. Brass and winds—well, that's a whole other enchilada.

Figure 9-2. Orchestra session at Ocean Way, Nashville, Tennessee, located in a church.

I use strings in both forward and background roles, creating lush landscapes if the song calls for it. Sometimes I'll work off of existing charts, and sometimes I'll create parts on the spot. Instrumentalists who can translate your ideas into string or horn parts on command are a rarity, so I stay very friendly with those special session musicians.

I like bendy, angular string arrangements, big swoopy unisons, soaring melodies, chugging deep rhythms. I like bright, stabbing horn parts and sexy saxophones. It is a disappointment if my arrangements are not received well, and I have to remind myself that clients don't always accept our ideas and visions. Just the way it is. Check your ego at the door.

Figure 9-3. Tracking with big strings.

Julian Colbeck on Playing Piano with an Orchestra

"When you are a rock-'n'-roll player, playing with people who are classically trained is a very humbling thing. It is tremendously exciting and terrifying at the same time. As a piano player, just being in the studio with the red light on can be frightening enough, but when you are doing it with a live sixty-piece orchestra, the pressure is taken up so many more notches. It's like being on a jet aircraft. You have this big thing surrounding you, the sound is loud, and the exhilaration is just phenomenal."

Geoff Emerick on "Depth of Field"

"As is the case when working in photography, you need to incorporate the depth of field. Today there is the danger of having everything in focus. There seems to be nothing out of focus. So it's important to try to create that depth of field. Create that little shimmer in the distance. The background being out of focus came about because of the only way we could work, which was with the rhythm section and the strings and brass and woodwinds, all playing together. Because of that, you've got a bit of the drums picking up on the string mics and a bit of the strings picking up on the woodwind mics and different equalization—you get all this shimmery stuff in the background."

Ed Stasium on Horn Poisoning

"It's got to be in tune, or it makes me queasy. One time I actually got physically ill from a horn section being out of tune. I had to go throw up. It might have been food poisoning, but I claim it was the horn section. And get this: The name of the band was Sun Rise Some Don't. Really bad name really bad horn section."

Figure 9-4. Homemade Decca Tree.

DECCA TREE

Al Schmitt is a guy who has recorded the greats: Tony Bennett, Frank Sinatra, Henry Mancini, Quincy Jones, Ray Charles—big, big sessions with big, big orchestras. He told me about a simple miking technique called a Decca Tree that he uses to capture an entire orchestra. I had to try one, and it didn't take very long to build our own custom Decca Tree contraption out of surplus mic-stand parts.

Per Al's instructions, we put three omni condenser mics on the three points of the tree and had it suspended over the conductor's position. I used Mojave MA-100 mics

Figure 9-5. Jefferson Baroque Orchestra with Decca Tree miking.

for the job. I first used the Decca Tree while recording a baroque orchestra, bringing up the three mics left, center, right. Wow, wow, wow. That's all that was needed. With a few little spot mics, everything was pretty much right there. That's it.

I started using this same technique for recording string quartets. Beautiful. Recording acoustic guitar? Also amazing with the Decca Tree. Backing vocals, acoustic piano, horn sec-

Figure 9-6. John Lennon's *Menlove Ave.* album.

Shelly Yakus on Phil Spector String Sounds

"On the Phil Spector string sessions for John Lennon, we would use a pair of mics as an overview for all the strings. Then there were some lower mics that were closer to the violins, that would pick up two or three at the same time over each of the sections. Then we would have a mic on each viola and a mic on each cello. Generally, strings sound better when miked farther away, because you don't hear the rosin from the bows, you hear the tone of the instrument. But for the violas, you need a good quality closer mic on them, because they don't speak as loudly as a violin would speak. And you also didn't want to get the leakage from the violins into the viola mics, so closer was better. But part of that big Phil Spector sound is leakage. Light travels very fast from its origin, but sound is very slow. The distance from the violins to the violas and the violins to the cellos creates enough leakage to make it sound really big! That's one of the things I learned early on: Leakage can be your friend, as long as it is good-quality leakage. If it's good leakage, it can really make something increase in size and sound really emotional."

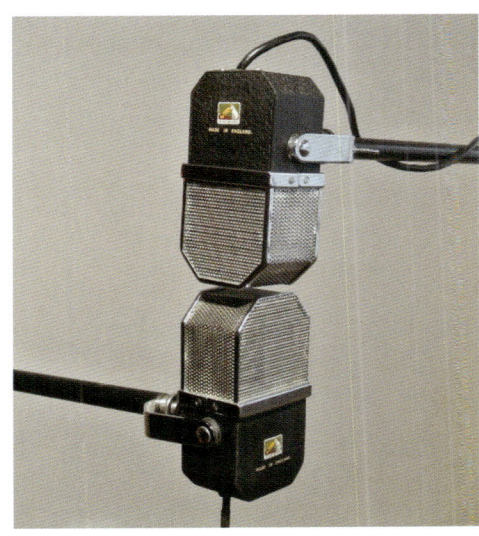

Figure 9-7. HMV classic ribbon mics.

Figure 9-8. De Geer's tube mics.

tions, accordion, percussion, vibes, drum kit overheads. My life is changed. The Decca Tree stays right where it is, in the middle of the tracking space. Permanent fixture. Thank you, Al.

ULTIMATE MICS

Let's explore microphones for strings, horns, reeds, wind, and orchestra. For capturing orchestra, these custom-built Didrik De Geer tube mics are my ultimate choice. They are astonishing. Numbers 0007 and 0008 are rentable, and this may be the only way to use them, as Didrik is retooling his "factory" (it is truly a one-man show) and the mics may not be back into production for a few years. Be looking for them! They are very special!

Figure 9-9. Royer R-121 ribbon on cello.

For strings and horns, ribbons are king. They round out the screechiness and harshness of those sounds, settling them comfortably in a sonic painting. Royer makes durable, wonderful-sounding ribbon mics for modern recording. The Royer R-122V combines the sound of a rib-

STRINGS, HORNS, AND ORCHESTRA 157

bon with 15 dB more gain than typical ribbons, giving you the ability to use this mic with most mic pres.

Sure, I love the ribbons and the tubes, but I've also just slapped a Shure SM57 up there on a violin. In fact, I horrified a Russian string quartet by recording them with $100 mics! And you know what? It sounded just fine. Be not afraid!

Alan Meyerson on *Black Hawk Down*'s Confrontational Horns

"On the Black Hawk Down film scoring sessions, composer Hans Zimmer and I wanted to record the music in a 'confrontational' setting. We went with all brass, and instead of having them arranged in the studio like a traditional orchestra, I set up the players in two rows so they were facing each other. Down the middle I placed a row of Royer ribbon mics, which are bidirectional. This way, they would play at each other in a competitive way."

Eric Valentine on Using Ribbons on Everything

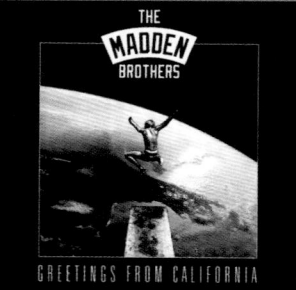

Figure 9-10. The Madden Brothers' *Greetings from California* album.

"Some of these old ribbon mics can still be gotten for a fair price, and they sound incredible if you get the right ones—Shure 300 ribbon mics, Reslo ribbon mics. On these older ribbon mics, because they all sound different, I'll buy several, and eventually you'll start to arrive on the ones that sound the best. The newer Coles 4038s are great for guitar room, but they don't handle close miking. I have a vintage RCA 44 that is actually one of my favorite mics. I actually tried an experiment where I recorded every single instrument with that microphone on a song, just to see what it would happen. And it worked great! The song was 'Dear Jane' by the Madden Brothers. I used the RCA 44 as the primary mono drum-room mic. I used it for all the guitars, used it for all the vocals. I used it for every single component of this song. It was all overdubbed one thing at a time."

Hans Zimmer on the Quality of an Instrument

"Aleksey Igudesman is my main violinist. He has one of those sort of violins—it is actually not his; it belongs to a bank. It's worth $9 million or something like that. What is so interesting about it is, when he plays one note quietly, it's twice as loud as the Roma gypsies' violins from Eastern Europe. Because the gypsies' violins are so cheaply made, it influences the Roma's style. They play harder."

Figure 9-11. Strings on Pauley Perrette's album with Suzie Katayama at the Village.

George Massenburg on Earth, Wind & Fire's Horns

"I started out as a tenor trombone player who was also interested in electronics. By the time I went to record Earth, Wind & Fire at Caribou Ranch, I had already learned the ranges of all the brass instruments, and knew what out-of-tune, sloppy playing sounded like. So when I was running horn overdubs I was fearless. If I wanted another take, I'd usually pretend there was some engineering problem where we would need to take another pass at an overdubbed horn part, often making these poor trumpet players play pass after pass until it was played in tune and on time. Pretty impressive just how much a great live performance improves a recording! By the way, I played the same role when we were recording vocals."

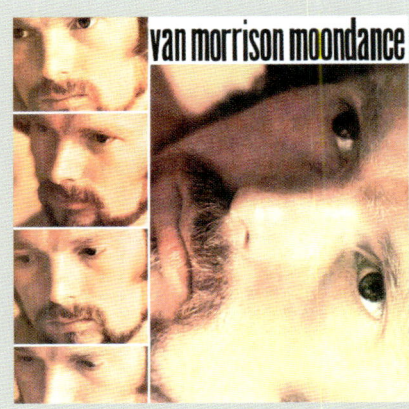

Shelly Yakus on Van Morrison's *Moondance* Horns

"One of my first solo engineering gigs was Van Morrison *Moondance*. I knew if I screwed it up, they'd never call me back in again for another engineering gig, but I was ready. I used a Neumann U 87 for the trumpet because that's what I saw Phil Ramone use, and a Sony C37 mic for the sax. Elliot Scheiner and myself wound up doing the whole album."

Figure 9-12. Van Morrison's *Moondance* album.

Geoff Emerick on Getting a Horn Blend

"On the trombones and trumpets, you can have them playing together, using the 4038 Coles ribbon mic because of its figure-eight pattern, with trumpets on one side of the mic and trombones on the other. Then they balance themselves and you can record a good blend."

Alan Meyerson on Brass Choreography

"Instead of the normal 'conducting' of musicians from prewritten charts, orchestrator Bruce Fowler (Mothers of Invention) would create a lexicon of certain arm movements, relating them to particular actions. So one arm movement would mean 'rise and get loud,' another would mean 'rise and get soft,' and so on. On the *Black Hawk Down* scoring sessions with Hans Zimmer, Bruce would get up there on the podium and wave his arms like a madman, choreographing the orchestra, creating textures and sounds. Rhythmic stuff—horns going *babababa* like machine guns. French horns dive-bombing like jets. Then one of us had this idea: What if the horn players didn't have any particular place to stand? So we told them to just walk around the studio while we recorded them. And they did it, marching around in no particular order. Poor horn players. If you asked a string player to do something like this, they'd think you were crazy—but brass players just have a beer and do it!"

Figure 9-13. Morgan O'Shaughnessy demonstrates how to play a nyckelharpa.

AND WHAT ELSE?

Beyond the traditional, there is a whole world of exotic international instruments. Never hesitate to drag home a strange foreign noisemaker when you find it hiding in a shop corner. My friend Jack Schumann would collect an unusual instrument from every place he visited. Over the span of sixty years, he found nyckelharpas, tamburs, elephant bells, biwas, and pipas. Eventually, people began collecting for him, sending him discoveries from all over the world. Today, the seven hundred-plus instruments of the Schumann Collection live at Southern Oregon University, there to be experienced by a new generation of musicians and producers.

Figure 9-14. Jack Schumann with his impressive collection.

Nick Launay on Johnny Rotten's Violumpet

Figure 9-15. Johnny (Rotten) Lydon with Steve Jones.

"John Lydon came in and asked, 'What 'ave you fuckers been up to? 'ave you done anything useful?' So I played him everything we had worked on that day, and he'd say, 'Well, that's fucking rubbish.' But when we played him this one drum track, he immediately liked it, and he said, 'Well, that's a bit all right, idn't it?' And then he just said, ''ave you got a mic? Let me hear it again.' And he listened to it twice, and then he ripped open a packet of cigarettes, and he used the inside of the packet of cigarettes to write on—I mean, there was paper available, but he just did that—and he wrote the lyrics to this song, and he instantly went out and did the vocals. And we added only one other instrument, which was a violumpet. A violumpet is a violin that has a trumpet sticking out of the top of it. And he just happened to have one—I think he bought it at a secondhand shop or at a flea market on the way in.

"You put the violin part on your shoulder, and the horn goes round your neck. So he played it, and it doesn't sound like a violin, it sounds like an Indian snake-charmer instrument. So we put that on the song, and I flipped the recording backwards. Well, I thought it was already quite good, but it sounded even better backwards. The song is 'Four Enclosed Walls' on Public Image Ltd's *Flowers of Romance*."

Figure 9-16. Stroh Violin, or "violumpet."

Al Schmitt on Paul Horn's Bagpipes

"When I was a producer at RCA, I did an album with Paul Horn where we had five bagpipes playing—on a jazz album! Well, bagpipes will drive you nuts, because they never stop. They've got to be blown up to get the sound going, and once they get the air set and they're making noise, it just doesn't end—you know it will eventually die out, but it takes a long, long time—so for three hours, it was pretty obnoxious. And it was all recorded live!"

Ed Stasium and the Alpenhorn

"There was a brother-and-sister group called Sihasin, a Navajo band that I recorded in Durango, Colorado. While doing their album, they brought in a friend, Werner Erb, a fellow who is known for his work on the alpenhorn. So yes, we had him play on the record. The thing was huge. When you play the alpenhorn, it kind of goes *whoooooooooooer*, in big, long notes. You can't really jam on alpenhorn, so I recorded a bunch of samples that I laid into Sihasin's music in different layers with delays on it. I use Audio-Technica mics almost exclusively, so for the horn recording I would have certainly used AT mics."

Figure 9-17. Ed Stasium and Werner Erb with the Alpenhorn.

Lori Castro on Joanna Newsom's Harp with Dog

"I assisted the sessions for Joanna's *Have One on Me* at RadioStar in Weed, California. Fantastic material, everything was recorded to tape. Actually, a dog from a neighboring session snuck in one day. It crawled all the way up to the harp during a take. You can hear the jingling on the playback. The dog sniffed all the floor mics and then her shoes. And after the take, Joanna said, 'That is awesome! This song is about a rabbit, so this works out great! Let's keep it!'"

Figure 9-18. Joanna Newsom recording her epic harp for *Have One on Me*.

Hans Zimmer on *Gladiator*'s Duduk

"On the film *Gladiator*, I was writing for this great Armenian duduk player, Djivan Gasparyan. I heard his music, and I started writing for him to be on the score, even though he did not live in the States and I did not have a contact for him. I'm very close to the filmmakers, so it makes experimentation and communication a lot easier, so I kept writing these pieces for Djivan anyway, and the film producers were getting worried. 'You're not going to Armenia to record this Djivan, are you? And he's not going to come here?' And director Ridley Scott would say, 'No, no, let Hans do this.' Finally, there was a day when Ridley came in to say, 'Look, Hans, I think you are going to have to get serious about this, because we've got a movie to finish! I think you need to give up on the Djivan idea.' And within twenty minutes of Ridley showing the tiniest amount of doubt, while he had been defending this idea all along, I got this phone call from guitar player Michael Brook, going, 'Oh, Djivan is coming to L.A. and I'm going to do this tour with him, and I'll be practicing with him for two weeks.' And poor Michael, I just stole him. And so Djivan came into the studio and didn't speak a word of English, and for a whole week we just made music. We could never say anything more than 'hello' and 'good-bye' and 'thank you.' After a week, we had all this fantastic music, and it never occurred to us that we didn't speak the same language."

Figure 9-19. Ross Robinson.

Ross Robinson on Korn's Bagpipes

"I did the bagpipes on the first Korn record outside at Indigo Ranch Studio. What you hear on the record was all done live in one shot. I had Jonathan Davis start from the top of the driveway, walking towards the studio. The band started playing. Jonathan was still playing the bagpipes as he walked into the room and went up to the mic and sang. Band still playing the song. He got to the mic with perfect timing. It was awesome. We had mics set up in that whole area, bagpipes really bouncing off the trees and the building."

Figure 10-1.

10

KEYS, SYNTHS, AND SAMPLERS

Figure 10-2. An orchestrion.

EARLY MECHANICAL BEASTS

The late 1800s saw the rise of the "mechanical revolution," when hand-crank-run, roll-driven music boxes and full-blown mechanical "orchestras" took hold in Europe and throughout the United States. Many of the most elaborate devices, known as "orchestrions," were developed in Germany and featured not only roll-driven player pianos, but also mechanically operated horns, reeds, pipes, flutes, drums, and even violins in the same large wood-and-glass case!

Original orchestrions were driven by a series of pneumatic relays, keyed off of wind-up perforated paper rolls, operated in the same way as the much-loved player pianos. Incredibly, paper-roll reproductions for the most popular antique units, including the smaller hand-cranked "organettes," are currently being manufactured by orchestrion enthusiasts. And new composi-

tions can be written for the antique pneumatic units, with the help of the roll-makers. Can you imagine writing original music for orchestrion?

Figure 10-3. Pat Metheny and the "Orchestrion Project."

Nick Launay on Gotye at Audities

Figure 10-4. Gotye.

"I know a fellow named Wally (better known as Gotye), one of the most phenomenal songwriters and musicians I've ever met and a lovely guy. He's Australian, and he likes a few Australian records I've made. So we met. And we became friends, and then he rings me up a month or two later and says that there is this museum in Calgary, Alberta, in Canada which has the biggest collection of synthesizers and keyboards in the world. And they've offered for him to record there.

"So we went up there for ten days, and we had a mobile Pro Tools rig, and we went 'round the museum with him playing and singing. And we did eighty hours of recording! He has a photographic memory, basically, where he can be told something once and it goes in, stays in, and he understands it. So when the fellow there at the museum would say, 'Oh, this works this way and you do this,' he got it. And Gotye could then play it as if he had played it before. It was great having this Pro Tools rig on wheels, and each day we'd just move it a few more feet and record something else. They had this massive orchestrion at the museum. It took half the space and Wally, being as prolific as he is, composed a whole piece on it while we were there."

Well someone did. Jazz guitarist Pat Metheny has created an entirely new mechanical format, based on MIDI / control-voltage-driven solenoids, operating a massive room-size instrument, which he calls the "The Orchestrion Project." This massive instrument has drums, robot guitars, vibes and bells, cymbals and traps, bass, and a completely unique "bottle organ" designed by Peterson (a company known for its tuner design).

The Early Synths—Rediscovered

Synth technology has obviously "improved" since the birth of the theremin. Today's presets are a distillation of fifty years of tinkering to find a useful sound. Maybe that is enough—or maybe not! Dig back into the historic instruments and redesign those sounds. Don't conform to someone else's ideas! Redraft the army of analog synths that started the synthesizer revolution.

Figure 10-5. EMS VCS 3 synth, from the collection of Brad Wood.

Hans Zimmer on Being a Synth "Composer"

"In the late '70s, early '80s, I was the Michael Boddicker of London. I was the guy who had a Prophet-5—serial number 26, (so you know it never worked)—and I had an EMS VCS suite and a Moog modular, so I was known as the 'guy who could make sounds.' I was an appalling player, and to this day I am an appalling player, but I was always good at making sounds! I'll sit there and start making sounds, and days and days and days will go by, and I'm still tinkering on my sounds, as opposed to writing notes! I became a composer by luck.

"When composing for pure synthesizer music, I don't write parts for 'strings' or 'horns.' We will start with a tune, but sometimes the sounds just take you in a different direction. It's like you are an artist, but instead you are painting with sounds. Let's just forget about strings and normal orchestra and go wild here."

Figure 10-6. Hans Zimmer in his massive synth lab.

Julian Colbeck on the Most Disgusting Polymoog

"I am very well known for hating the Polymoog. I loathe that thing with a vengeance. But I had to play it when it first came out. It was one of the first polyphonic synthesizers. As a keyboard player, you might have a Hammond. If you were wealthy, you might have had a Mellotron, or a Clavinet. But as soon as synthesizers came in, they became the Holy Grail. It started as monophonic only. Polyphony was this dream—'Oh my god, one day we'll be able to play more than one note at the same time.' And one of the first—if not *the* first to do it—was the Polymoog keyboard. Which is an absolute dog. To me it was one of the most disgusting Moogs I've ever heard. I was playing with an English pop star named John Miles when it first came out. We did a big TV show in Munich. Fifteen minutes before taping and they gave me this Polymoog, and I couldn't believe it. Just the most horrible thing and unplayable."

Figure 10-7. Polymoog, from the collection of Sweetwater's Chuck Surack.

Damon Fox on his Moog Modular and Not Being Cool

Figure 10-8. The Moog Modular, from the collection of Damon Fox.

"In the '90s, analog synthesis was a dead art; it had really fallen out of favor. It wasn't really cool to have an old synthesizer. And here I was hauling around my Moog Modular to shows. When we were on tour with my band Bigelf, we would actually take the Moog Modular into the Denny's with us when we ate, because we were afraid that it would get stolen in the van. And it was huge! The Modular is something so special. You can get these truly mechanical sounds that are raw, big, organic, and real. Not glossed over like a softsynth or a plug-in."

Jeffrey Lorber on Redesigning U2's "Desire"

"Taavi Mote and I were hired to work on remixes for the U2 *Rattle and Hum* album. The first song we did was 'Desire.' It was very unusual those days for a band to come to a remix session, but all four guys came down, which was awesome ('Where's the Guinness?'). Bono was so excited about the powerful groove we were creating that he wrote a few new verses right on the spot. The Edge was very interested in the fact that I doubled the bass line with a Minimoog and an acoustic piano part. After a while, we also decided to get some 'cinema verité' to add to the song, so we recorded a car alarm and random stuff from the news right off the speaker on the TV above the console in Larrabee Studio B. Those elements ended up being featured prominently in the mix and were dramatized in the video they made for the song."

Julian Colbeck on the ARP Odyssey

Figure 10-9. Julian Colbeck.

"I was off to buy my first synthesizer. At the time, it was either going to be a Minimoog or an ARP—those were my choices. I went into a shop at Charing Cross Road in London, and the head salesman of the keyboard department was a guy named Dave Simmons (who later developed Simmons Drums). We spent around an hour fiddling around with this ARP Odyssey, and neither of us could get any sound out of it whatsoever. Pushing and pulling all these sliders up and down and getting nothing. These were very early days, and neither of us knew what we were doing, and the Odyssey was a very strange thing. Anyway, God knows why I bought it, because to me it always sounded horrible. In hindsight, I've always wished that I'd bought the Moog instead. It was such a superior instrument."

Figure 10-10. The DK Synergy II+, from the collection of Damon Fox.

Damon Fox on the Frustrating Synergy

"The DK Synergy II+ was an early-'80s additive synthesizer thing. It's super temperamental and pisses me off constantly—and it's all over my new record. It takes these cartridges, and it actually comes with a Kaypro, an old computer with a green screen so you can draw the waves. Complete FM synthesis. It has a 'retro-future' sound. I think of it as the 'steampunk' of synthesizers. Wendy Carlos used something like this on the original *Tron* soundtrack. Really cool stuff, but totally frustrating. Right in the middle of using it, the screen will go red and the thing will just spin out. The sound will go wacky, and I'll scramble to record these random, weird digital noises, so it always has me running to hit the space bar to try to capture the insanity. I think a capacitor is going out on it, and there is only a short window of time when I first turn it on that it works properly. After that, good luck. No, I won't be getting it repaired. Of course there will be a sound that I'll miss, and I'll want it to do it again, and it will never do it. I guess that's part of the fun of it!"

Figure 10-11. Sound cartridges for the DK Synergy.

PATCHING THROUGH ANALOG SYNTHS

When you are using an analog synthesizer, the oscillator is the thing making the actual tone note. After the oscillator, the tone will run through an ADSR (attack, decay, sustain, release), which is an envelope filter to shape the sound of the note. Some synths also have high-pass EQ filters, spring reverbs, and other cool effects the generated tone can run through. On the ARP

Figure 10-12.
The ARP 2600.

2600, the EMS VCS 3, the Moog Modular and the Korg MS-20, you can patch in between the oscillator and the ADSR. This is cool because you can affect whatever sound you patch into it as you would the synthesizer-generated tone—so you can make a vocal go "wow," or a piano sound like a radio drenched in springy reverb. I used to bring my ARP 2600 to sessions and create all kinds of unusual effects by using the synth as a filter in this way.

Larry Crane on Organic Synth Pulses

"I love to trigger gates to open a synth keyed off of different tracks. I'll use an analog synth like a Moog, playing and holding down the bass root notes in the song, running it through a real analog rack-mount gate like a Drawmer DS501. Then I run a kick and snare to key the sidechain of the gate. The drums open the gate, and you can hear the synth note come through for the duration of the drum hit. Or I put the drum keys into a delay and have it repeat, opening the gate so it pulses with the music. It creates this cool New Order kind of synthesizer sound without being as clinical and mechanical as a sequenced track."

KEYS, SYNTHS, AND SAMPLERS

TAPE- AND DISC-BASED SAMPLERS

Early samplers are fascinating to me. My curiosity about early tape-based keyboards drew me into purchasing Mellotrons and Chamberlins, and overusing them on a decade's worth of my productions. But the challenge is lost when you use a modern sample of one of these old devices. For one thing, the modern sample is in tune! All the keys work! There are no garbles or clicks or fuzzies! When playing my original old Mellotron 400 keyboard, I had to move my fingers in a certain way, in a certain order, with certain keys requiring a particular touch to get them to work. It was fun and exciting to work with these somewhat difficult tools. And so rewarding.

Figure 10-13. Josh Dies from Showbread with the Mellotron M400.

Optigans use clear disc-based systems for delivering the crustiest-sounding crude samples of lead instruments with bass and rhythm accompaniment. Each disc has a theme: Polynesian, Latin, country guitar, and, my favorite—the "Banjo Sing-A-Long." I find these weird plasticky keyboards in thrift stores. They look like the cheapest of kids' toys from the '70s, and they are frickin' cool.

Figure 10-14. The Optigan and its clear discs.

Figure 10-15. Damon Fox and his 360 Systems.

Damon Fox on One-Upping Roger Linn

"I've seen this 360 Systems digital polysynth in an ad in *Keyboard* magazine from 1983 for $4,995. It's basically the '80s version of the Mellotron. From what I was told, the guy that made the chips for the white boxes at the airport that said, 'This zone is for the immediate loading and unloading of passengers,' well, legend has it that the chip guy and Roger Linn used to go to lunch together. Roger told him that he made drum machines so chip guy decided he could one-up Linn, so he made this keyboard. It's got a super-cool sound, and I bought it in a junk shop for ninety bucks."

THE WEIRDER, THE BETTER

There is a new generation of analog synths. There is the Swarmatron, the Knifonium, the Doepfer modular, and Moog has a bunch of reissues. Korg also has many new analog synths,

Figure 10-16. Sublime's Eric Wilson and his Swarmatron.

including the amazing MS-20 and littleBits analog synth kits and modular systems that you can build yourself! Moog has the Werkstatt synth kit to get you started. Let's do it all again, right back at the very beginning!

JULIAN COLBECK ON MEMORY "FOSSILIZATION"

"Reissue Moog Modulars are again unpredictable and unrepeatable. The sound has not been fossilized like it has been in modern samplers and emulators with presets, so if the knob is slightly to the right or left, it will sound different than the time before. And the whole point is that the sound would be for one time only. You are producing the energy at that moment. When memories first appeared on the Sequential Circuits Prophet-5 in 1978, I thought, *My god, finally we can capture these things*, and I think everybody was very starstruck by that capability at the time. Years later, I think that would become a problem with presets, where everybody sounds the same. You realize now that the memory was a false benefit. Actually creating something is always going to be more powerful than re-creating something."

Figure 10-17. Hans Zimmer's Knifonium.

Hans Zimmer on Heavy Lifting

"We actually dusted off the [Yamaha] CS-80 and tried to get it working. But while the CS-80 was still on life support, I thought, *Well, I'd better go and emulate the synth and work up a sound*. That's one of the problems, I have deadlines, so the stuff better work. There is, however, a plug-in I use called Zebra, which this rather ingenious man in Berlin named Urs Heckmann makes, so I used the Zebra in the box for a while. And in some time, we got the CS-80 up and working, but to be honest, the plug-in actually sounded better! If it's old or if it's new, it all serves a purpose. Everything has its place, though I have to apologize to whoever it was that had to lift the CS-80."

Figure 10-18. Hans Zimmer's Yamaha CS-80 beast.

Julian Colbeck on Capturing Bill Bruford's MIDI "DNA"

Figure 10-19. Bill Bruford.

"The cool thing about MIDI recording is that it actually captures the 'DNA' of the player, not just capturing the sound. So you're getting 'note on,' 'duration,' 'velocity,' 'note off.' And by doing that, you can really see what the music is, much more than you can ever see in an audio recording. I recorded Bill Bruford for a project I called 'Twiddly Bits,' which is a set of MIDI-based loop recordings that you could insert your own sounds into.

"I had worked with Bill in Yes for many years, and he was using a fairly complicated instrument called an SDX, a very high-end sampling drum kit at this time. So with his MIDI recording, you could play back a Bill Bruford groove on a Roland 808 kit, a jazz kit, a hip-hop kit, anything. You could even apply bird chirps and glass-clinking samples into the groove and, without any question or doubt, know it was Bill Bruford. The sound made no difference at all! With Bill, his kick drum is always slightly behind, his snare is always slightly ahead, and you can see his body clock. And that's in the MIDI! You'd never see that in an audio waveform. MIDI actually lets you analyze what the player is doing. To me that is their 'DNA.'"

Figure 10-20. Justin Stanley recording on Doug Aitken's "Station to Station."

Figure 11-1.

11
PERCUSSION AND OTHER NOISE

SLAPS, DINGS, AND RATTLES

Percussion is the sound of hitting or scraping one object against another. That is the official definition of the word. So, when percussion is viewed as simply as that, anything can be percussion! Jars and bottles, hammers on anvils, brooms on sidewalks, wrenches on trunk hoods, ice cubes in a glass, keys, doorbells, impact hammers, fingernails on blackboards, woodpeckers, coins, silverware, pebbles, seashells, peanuts, vitamins in a bottle, typewriters, staplers, sugar packets, prison doors, floor jacks, railroad tracks.

Figure 11-2. Pete Thomas uses frozen trout as percussion for P. J. Harvey.

Figure 11-3. The Shure Beta 181, a mic that is ready for anything thrown at it, with four interchangeable capsules!

Brian Wilson used empty Coca-Cola cans and trash cans as percussion on the Beach Boys' *Pet Sounds* album. Why does anyone ever reach for a tambourine when your kitchen is full of jangly percussion instruments?

MICHAEL BEINHORN ON THE SOUNDGARDEN SPOONS

"'Spoonman' was the first single off the *Superunknown* record I produced with Soundgarden, and had been written about a Seattle street musician named Artis. Artis was something of a virtuoso on the spoons and, as we discovered, his performances were the stuff of legends.

"The day we were ready to record Artis, he showed up at the studio with what looked like an oversize bedroll. He brought it into the recording area, set it on the floor, and opened it. Inside was an extensive collection of battered spoons, along with some bells and many flat pieces of metal of various sizes and shapes. Jason

Figure 11-4. Artis the Spoonman.

Corsaro put up a pair of ambient U 67s to capture the performance. Shortly before we got started, Artis announced to us that we needed to have a video camera on hand because, as he put it, 'You're gonna want to get this on film.' And so saying, he removed his shirt. With that kind of build-up, I was ready for something interesting, but I had no idea what was coming. Although the song had a specific section for the spoons to solo, we rolled the track from the top, and, so to speak, let him rip.

"And rip he did. From the moment the track began, Artis began grabbing spoons, odd bits of ringing metal, and anything else he could find in that bedroll of his and slapped them against his bare skin as hard and as fast as he could. At one point I looked around at everyone else in the control room. They were all gaping in amazement—wide-eyed and slack-jawed at what was taking place. The spectacle and performance were so intense that it felt as if the whole thing could fly off the rails at any instant, but it never did. This guy was literally beating the shit out of himself with these bits of metal at almost blinding speed, running them up and down his body, while playing these complex rhythmic subdivisions, all executed with preternaturally perfect timing.

"And blood was flying everywhere. What had started out as a recording session had quickly turned into a piece of sadomasochistic psychodramatic performance art. The whole thing took Artis four passes. Between the cardio, the intense exertion, and the loss of blood, a lesser man would easily have passed out a few bars into the first pass. That afternoon, he gave a whole new meaning to the term 'suffering for one's art.'

"By the time Artis had finished, we were all kind of devastated—stunned into silence and a little scarred by what we had just witnessed. By contrast, Artis was completely pumped up and loaded with adrenaline from his previous thirty or so minutes of flailing, gyrating, and bloodletting. He came bouncing into the control room with his bedroll under an arm, and for the next hour proceeded to regale us with his philosophy regarding virtually everything—which had clearly been inspired by a massive infusion of hallucinogenics at an earlier stage of his life. And, in true street-musician form, he took the advantage of his captive audience and attempted to sell a book full of his musings and poetry. That was a truly unforgettable afternoon."

Figure 11-5. Soundgarden's *Superunknown* album.

Ross Hogarth on Keltner's Unique Shakers

"Back in the day, Jim Keltner had bags of these pot seeds that he'd put in different kinds of canisters. I guess they were left over from cleaning his Mexican weed. He would put these seeds into metal film canisters, different kinds of bottles, creating really unique-sounding percussion."

Figure 11-6. Jim Keltner always protects his eyes against cymbal glare.

Tim Palmer on Pearl Jam's Kitchen Percussion

Figure 11-7. Pearl Jam's *Ten* album.

"Necessities are often the 'mother of all invention,' and in the recording of music, these necessities can often lead to a more original and creative sound. Stretching the possibilities is definitely a good thing. People often ask me how I ended up getting a credit on Pearl Jam's *Ten* Album for playing 'pepper shaker and fire extinguisher.' Well, that's a perfect example. We were mixing the *Ten* album in the beautiful countryside of Surrey, England. The studio was a converted farm. We had pretty much everything we needed, but music shops were not close by, rentals even further. I wanted to add some movement to the track 'Oceans,' and a trip to the kitchen fixed the issue quickly and perfectly. I grabbed a fire extinguisher and a container of black peppercorns. The fire extinguisher worked perfectly being played with a drumstick as a kind of metallic bell sound, and the pepper was a perfect shaker."

Figure 11-8. Mark Pauline's "Spark-shooter."

THE ART OF PANDEMONIUM

San Francisco mechanical artist Mark Pauline was the godfather of the industrial art movement. With his Survival Research Labs, he put on shows with giant mechanical monsters that would breathe fire and throw eggs at each other. Then they would dismantle each other, throwing sparks over the crowd, generating impossibly offensive clatter and roar. This was the

Michael Franti on the Beatnigs

"Our politics in the Beatnigs were very antiestablishment, but we did it through joy. It was a super-fun time of experimentation. We pounded on large pieces of metal as I did my poetry over the top of it. We had a huge metal swing-set sculpture that had tuned sheets of steel suspended by chains from a tubular frame. It sounded like bells. And we built other instruments. Garage door springs stretched between two pieces of wood with contact mics on them. It sounded like thunder! Rono Tse would use tire chains and smash them against huge pieces of metal. He'd use tire rims as marching-band-style cymbals. There was nothing else like it at the time; we were making it up as we went along, inspired by Black America, hip-hop, and roots poetry. We liked the Watts Prophets, Last Poets, reggae; and we were equally inspired by bands like Ministry, Joy Division, and the 4AD Records artists."

Figure 11-9. Michael Franti.

inspiration for industrial poets the Beatnigs, who used power tools and hammers on steel to create a sonic spectacle. Fronted by Michael Franti, the Beatnigs laid the foundation for the Disposable Heroes of Hiphoprisy and Franti's Spearhead projects.

I was "lucky" enough to be living in the Beatnigs' flat in the Mission District where they would rehearse. In the front room of the flat was the Beatnigs' huge metal percussion "swing set," with large, heavy cut pieces of plate iron, each fabricated to ring at a certain pitch. The metal was suspended by heavy-duty metal chains and was played with hammers: thundering drums, clanging steel, sparks flying. Needless to say, it was impossible to sleep in the flat when they were working on their set.

Figure 11-10. Nick Cave and Warren Ellis.

Nick Launay on Trashing Up the Birthday Party

"During the recording of the Birthday Party, Nick Cave's original band, I did a song called 'Release the Bats.' Needless to say, the song is probably about women with vampire teeth, biting. It's one of my favorite tracks of all time, and it has a crazy drum sound on it. It's very tom-based and it's, ah . . . very tribal. And then it hits the chorus, and the chorus is just brutal. Nick is just going, 'Release the bats, release the bats, release the bats!' 'Don't tell me it doesn't hurt.' 'Bat bite, bat bite.' We had all these guitars going through three or four EQs, and it still wasn't aggressive enough, so we brought sheets of corrugated iron into the studio and just bashed it. It just sounds like the drum sound gets very trashy, but what it is, is all of them out there just bashing the metal with drumsticks and whatever they had. When you hear it, that's very obvious that's what it is. It's just not what you would expect."

Figure 11-11. Joe Henry.

Joe Henry on Luggage Carousels

"For me, the overall mood of the track is as important as any instrument that is played. Working with engineer Husky Höskulds, we used a handheld cassette recorder to record the turning metal luggage carousel at JFK baggage claim, slowed it down, and ran it beneath an entire track of my album *Scar*—creating a groaning tension almost like distant weather."

Alan Meyerson on the Cosmic Beam

"During the scoring of the *Thin Red Line* film soundtrack, composer Hans Zimmer had me record this crazy instrument called a 'Cosmic Beam.' It is basically a huge construction I-beam strung with piano wire, with pickups built into it, feeding a 200,000-watt sound system. It's played with rubber mallets that you swipe across the strings, that creates this giant whooping sound. Well, I went to the warehouse where it lived to take a look and listen to it, and realized there was no way I'd be able to record this thing where it sat. It was way too big a sound for that space! So since it was for a Fox film, we thought to go over to one of the massive 50,000-square-foot shooting stages on the Fox movie lot to do the recording. And we did it! My job was to mike this puppy up and make it sound big. I was running around, climbing through the rafters, hanging mics everywhere. Every mic I was using had to take 150 db of sound levels, so they were DPA and B&K mics. In the movie scene where the Americans run up a hill to attack a Japanese camp, you hear this rumble, this thing that starts. What the hell is that? That's the Cosmic Beam!"

Ross Hogarth on Stan Lynch Face-Slapping

"So I asked Stan Lynch, 'Can you come up with a rhythm sound using your hands slapping your knees or chest?' And of course he did, and started slapping his face, too, so I miked up his face! He didn't have to hit himself hard, not like it made marks or anything, because y'know, those sounds are just kind of loud to begin with! A good microphone picks them up easily. So Stan came up with a great rhythm pattern on a song for the *Brides* album by playing his body, his chest, and his face!"

Figure 11-12. Belly percussion.

DOWNRIGHT WACK-A-DOO

I have on more than one occasion played a chicken solo. There is a certain technique to it. It's best to learn these techniques before attempting to play a chicken in your studio. OK, so with headphones on your head for monitoring, you hold the chicken in place so its beak is pointed toward the microphone. I prefer a small diaphragm condenser and have had great luck with the Mojave MA-100 mic. Then, when you hear the part coming up in the song, gently squeeze the chicken until it begins to squawk. "Bah-gawk!" You can control the amount of squawking by the amount of pressure you apply to the bird.

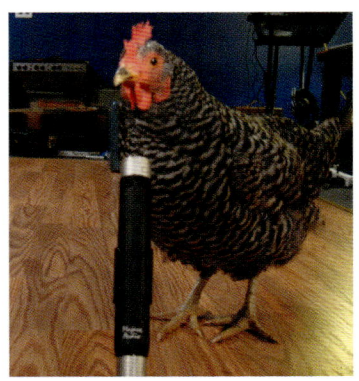

Figure 11-13. Chicken recording.

Figure 11-14. *How to Mike a Chicken.*

Figure 11-15. Sylvia playing a motorcycle solo.

PLAYING A MOTORCYCLE SOLO

Another great instrument for solo performances. You just have to find a way to get it into the studio and turn off the motorcycle between takes. Of course, you can record your motorcycle solo outside, but then that would be too easy!

PRODUCER GODS

Figure 12-1. Producer gods.

12

PRODUCTION APPROACH

WHAT DOES A PRODUCER DO?

The broad definition of a record producer is someone responsible for making sure a project gets done within time and budget constraints. Another way to look at it is that there are three sub-types of producers: the musician/producer, like the Neptunes or Rodney Jerkins;

Bob Ezrin on "Producing"

"I find the variety in producing music exhilarating and challenging, and a huge part of the joy of doing this for a living. We don't 'work' music. We 'play' it. And that's all that this is: one huge, ecumenical, apolitical, multinational and multicultural tent full of people playing. That's just about the best place in the entire world for creating beautiful things."

Figure 12-2. Bob Ezrin.

Figure 12-3. George Martin and the Beatles, circa 1966.

the engineer/producer like Jack Joseph Puig or Nigel Godrich; and the fan/producer, like Rick Rubin or Jimmy Iovine. The musician/producer usually creates all the music and often writes the songs, bringing in vocal talent to front the project. The engineer/producer will craft the sound of an existing project, often using equipment and technique to create the magic in the studio. The fan/producer may never actually touch the console, but will help choose the songs and guide the project by bringing the right people together.

My production style falls mainly between the engineer and the fan, but I'll often add

Al Schmitt on Producers, Bookies, and Hookers

"I would get these guys, producers, that would call all the time. They'd want me on their session—'We gotta have Al Schmitt, we gotta have Al Schmitt'—I would be the engineer, and they would be on the phone all the time with their bookies and hookers! They were the producers, but I'd wind up doing all the production on it. On the song 'What Kind of Fool Am I?' with Sammy Davis Jr., the producer never showed up! He forgot about the date! On another date I did with Connie Francis, we were doing four songs, and the producer was there for the first two songs and then left! That's when the conductor, Dave Rose, and I did 'My Happiness,' the third song. We came up with the idea to do double voice and bring another tape machine in—it was all mono back then. So we played back one machine back while she sang to it and recorded it on another machine—that's how we got the double voice. The producer was never even around!"

musical elements, including string arrangements, vocals, and Mellotron. My function in the studio is to keep the project going, to tell an artist when a performance is not good enough, and to let them know when it is. Many artists can't tell when they have finished, especially with endless Pro Tools tracks and the ability to continue to change without commitment. Someone has to say stop or the record never gets finished. So I can be the artist's reality check.

THE PIONEERS

Pioneering producers had the power to change the sound of music, accompanied by advances in technology that led them in new directions. They pushed further, built gadgets to express their ideas, and used recording as an art form and a message delivery system. Some of the most recognizable of the production pioneers include Les Paul, Sam Phillips, Tom Dowd, Phil Ramone, Teo Macero, and Willie Mitchell. Alan Lomax documented roots music, changing our perception of blues. Joe Meek created a whole genre of surf music with his "Telstar." Berry Gordy created a musical empire in Detroit with Motown. R. P. Weatherald created the concept album with Woody Guthrie's *Dust Bowl Ballads*. Brian Wilson turned the recording studio into his playground, and Sir George Martin translated the kooky ideas of the Beatles into magical, musical masterpieces. Here are a few other pioneers who made a big impression on my production style:

RAYMOND SCOTT

Raymond had a knack for creating a scene with music, putting you in a bustling factory with his famous "Powerhouse" theme, used in Warner Bros.' Looney Tunes cartoons. His '30s-era recordings had you swingin' with penguins at the North Pole. Or he'd put you in

Figure 12-4. Raymond Scott's Electronium, pictured with Mark Mothersbaugh of Devo.

the midst of a pack of hungry cannibals. Or partying in a haunted house. Or shaking with a Turkish belly dancer. Raymond Scott's productions with music were "visual." Raymond also recorded some of the first electronic music—in fact, he invented an electronic music synthesizer of sorts called the Electronium, which has been purchased and restored by Mark Mothersbaugh of Devo.

SPIKE JONES

Spike toured with the "Musical Depreciation Revue." His recordings included gargling, gunshots, birds chirping, harps, cartoon vocals, car horns, fart noises, tap dancing, fire alarms mixed with operatic singing, quacking brass, ripping, Hawaiian saws, and drills. Spike Jones blended comedy and music with the use of fantastic live sound effects.

JOHN CAGE

John's use of oscillators, radios, turntables, and early magnetic tape to create music started in the '30s. While his misuse of traditional instruments made him a recording pioneer as well as a creative genius, he was also often misunderstood. His innovations in "electroacoustic" music vastly predated others and inspired so many popular artists, including the Beatles. Some of his music could be considered "unlistenable," but to me it shows his childlike sense of adventure and devilish humor. This guy just cracks me up!

Figure 12-5. John Cage's *Early Electronic and Tape Music* album.

PHIL SPECTOR

I am obsessed with Phil Spector. The guy was totally out of his mind. He has had problems with social skills and the law, but wow, he sure made music that had a certain distinct sound—completely unique, deep, and blurry backgrounds. Cavernous. Accented with chimes, bells, and sparkling keys. He created the "Wall of Sound" by cramming dozens of musicians in a not-so-big room, shooting holes in the ceiling, holding them in there for days at a time, session players fearing for their lives. Forcing them to "get it right." And when it was right, it was *soooo* right! The performers would musically dance in perfect unison, dressed heavily in reflective space from the studio's echo chambers. Exhausting to listen to, once you know what went into it—terrifying and beautiful, all at the same time.

SYLVIA VANDERPOOL-ROBINSON

She started her rise to fame in 1972, releasing the song "Pillow Talk" under the name of Sylvia, and it became a smash hit. I remember hearing it on the radio with all the moans and groans. The song was pretty damn sexy in 1972 for a kid like me, but heck, she had the same name I did!

Later on, as disco ramped up, Donna Summer would use the same trick in "Love to

Figure 12-6. Sylvia Vanderpool-Robinson of Mickey & Sylvia.

Love You Baby," but Sylvia did it first. But what Sylvia Vanderpool-Robinson did next would shake the music world and put it upside down forever! She produced a song using samples nicked from Chic's "Good Times," and put a rapper over it! In 1979, this had never been done before. "Rapper's Delight," by the Sugarhill Gang, was released on Sylvia's own Sugarhill Records.

Immediately it caused a stir. First, because of the reuse of Chic's original recording as the base for the song. This was the first time sampling had been used in pop music, and people were getting all bent out of shape. Nile Rodgers, the producer behind the Chic song, apparently went ballistic. Lawsuits were filed, and Nile got his cut. After the dust settled, Nile became one of the biggest fans of "Rapper's Delight"!

KARLHEINZ STOCKHAUSEN

One of the most controversial composers of the 20th century, Stockhausen commissioned some unbelievable recording scenarios, which has put him on my favorites list. The most ridiculously awesome was the "Helicopter String Quartet," a piece of music he composed for four string players who would be lifted off the ground in four separate helicopters during the recording. Each player was listening to the other through headphones, and a sound

PRODUCTION APPROACH 191

technician and mics were in each helicopter. The recording of the four players had plenty of helicopter bleed, but the sound of the rotor blades was actually written into the composition, and it works (in a bizarre way)!

Figure 12-7. Karlheinz Stockhausen's "Helicopter String Quartet," with each helicopter containing a string player and an engineer.

Al Schmitt on "Uncle" Les Paul

"Les Paul was my uncle's best friend, so I called him 'Uncle Les' from the time I was a little kid till I grew up. All the way till he passed away, I'd see him. We would always reminisce about those days in New York. He would always mess my hair up, no matter what. When I was little, it was OK. But when I got older and he'd keep doing that, it would piss me off!"

Brian Malouf on Working with Quincy Jones

"I remember working with Quincy Jones, and we would do five, seven, twelve takes on a song with five musicians in a $3,000-a-day studio—probably a $10,000-a-day when all the charges were added up—and if he didn't love the best take, we'd scrap it three days later and do it all again."

Figure 12-8. Selfie with Quincy Jones. Wow. Yes, I'm a little freaked out.

Shelly Yakus on John Lennon's Drumstick Story

"During John Lennon's *Walls and Bridges* album, I remember driving to the studio and hearing the Beatles song 'All You Need Is Love' on the radio before the session. I tried not to bother John with a lot of questions about how the Beatles records were made, but I had to ask: 'John, I just heard 'All You Need Is Love' on the radio, and is that a Mellotron?' He said, 'No, man, it's the London Philharmonic Orchestra! Are you kidding? But I will tell you that there are no drums on that song.' I asked, 'How could that be? I hear drums!' He said, 'It's actually one of the Beatles holding an upright bass by the neck, squeezing the neck with the strings, and Ringo hitting the strings with drumsticks while someone else was shaking a tambourine.' And sure enough, when I got home and listened and muted one side of my stereo, I could hear it clear as day. The clack of the sticks hitting the strings sounded like a snare drum, but it's not a snare drum. For the Beatles, there were no rules. Whatever was in their heads, George Martin figured out how to get it out onto a record, and he did."

THE RISE OF THE INDEPENDENTS

There was a set way of doing things at the label houses in New York and L.A. and even in London up until the late '60s. Engineers had particular roles and a special set of rules; producers were much more business-oriented. It may have been because of labels' and studio owners' sheer cheapness—combined with musicians' widespread use of mind-altering drugs—that real studio creativity took hold in the '60s. When the staff engineers rebelled against being treated like truck drivers and went off to do their own projects as producers, that's when records got really interesting. Suddenly, there was a competition among the studio rats to see who could do something completely different, and the age of invention was born in time for the '70s, making it possible for artists like Pink Floyd, Genesis, and the Who—in that band's golden age—to take hold.

Some of the great engineer/producers of this era include Roger Nichols, Geoff Emerick, Alan Parsons, Glyn Johns, Al Schmitt, Tony Clark, Steve Lillywhite, David Hentschel, Kit Lambert, Mickie Most, Elliot Scheiner, Andrew Loog Oldham, Jack Douglas, Bob Ezrin, and Roy Thomas Baker.

Jack Joseph Puig on Glyn Johns's Fistfights

"Of all the bands I've worked with, the Black Crowes was the most scrapper. I called my mentor Glyn Johns and told him, 'I think I'm going to have to get in a fistfight with this singer.' Glyn went on to tell me of all the bands *he* got in fistfights with. He told me that he and the drummer of the Who, Keith Moon, would get into fistfights all the time. Glyn said it is not unusual for fights to break out when you are dealing with that kind of horsepower. These great players have an incredible amount of attitude inside them."

LEE "SCRATCH" PERRY

It may have been legal issues that forced Lee "Scratch" Perry to mix only instrumentals of popular reggae artists, using little splashes of the main vocals in his mixes. And those splashes he drenched in reverb and delay. Perfect for the enjoyment of a good rolled green. Sweet Lord. Dub reggae was created right there in Jamaica. This genre enlivened multiple generations of dub reggae enthusiasts and went on to inspire a million mixers in studios all over the world. Including me. Dub helped me to discover that I could use the mixing board like an instrument. Perry made the mixer a part of the band.

Figure 12-9. Lee "Scratch" Perry.

Elliot Scheiner on Taking the Leap into the Unknown

"The early '70s is when freelance engineering started. All the on-staff engineers in the studios had been working for a modest wage every week, and we had a profit-sharing plan, but the profit-sharing plan was losing money, so I was thinking, 'Boy, at this rate, I don't know how long I can work in this industry, but if I leave I'm not going to have anything.' So I was the first guy in New York to go freelance. And it was really scary, 'cause I had been working at Phil Ramone's studio, A&R, for almost five years. And now I was going to do something completely different, but I had clients booked into A&R, lined up for seven or eight months already. So what was going to happen now?"

Figure 12-10. Elliot Scheiner.

Ed Stasium on Roy Thomas Baker's Drum Room

"I saw Steve Gadd taping a wallet to the top of the snare drum for a drum treatment. It gave the snare a punchy sound and got rid of the ring. I thought, *Oh, so this is what you do!* So I started taping anything and everything to the top heads of the snare and toms to get that tight effect. Paper towels, Kotexes, toilet paper—it was kind of a thing in the '70s to have real dead-sounding drums. There were big pillows in the kick drum, but when I worked with Roy Thomas Baker on a Pilot record, he didn't want any of that. We had a drum booth at Le Studio Morin Heights, and before Roy showed up to the session, we set the drums up in the booth and taped down the heads to deaden them like usual. Roy said, 'Bring the drums out here to get some of this beautiful room sound.' He also told me to pull all the damping off the top heads, so off came whatever I had taped on there. I learned from Roy Thomas Baker to use room sound and ambient sound and to let the drums ring. To this day, I love big room sounds on drums, guitars—anything."

Al Schmitt on Going Solo

Figure 12-11. Jefferson Airplane's *After Bathing at Baxter's* album.

"I was doing Eddie Fisher in the afternoon and Jefferson Airplane at night, and I wasn't getting any sleep, 'cause I'd do Eddie from 2 to 5 p.m., and from eight till four or five in the morning I'd be in the studio with the Airplane. So I called my boss and I said, 'I need some help. You're going to have to get someone else to do Eddie Fisher. I can't do both of these.' And he said, 'Why not? Truck drivers do it!' I said, 'OK, get yourself a couple truck drivers. I quit!' And I gave him my two weeks' notice and left!

"Back in those days, I was making $22,500 bucks a year, doing eleven acts as a staff engineer and producer. So the Jefferson Airplane called me two weeks after I left and said, 'Al, we want you to produce our records.' The label wanted to send someone from RCA to produce, but they said they didn't want anyone from there. So RCA said, 'OK, you can hire an independent producer,' and they asked if I would like to do it, and I said, 'Yeah, of course.' So I did it, and my first royalty check was almost $50,000 for doing one act!"

Geoff Emerick on Cutting Ties with EMI on "Abbey Road"

"I had just become an independent engineer, walking out after seven or eight years working at EMI's Studios at Abbey Road. Two weeks after leaving EMI, I was working for Apple Records. EMI didn't like that much. And then I came back into their studios with the Beatles to record *Abbey Road*, and EMI said, 'No, he can't come back! He's independent.' So there was all this negotiation. Part of the negotiated agreements was that one of their engineers, Phil MacDonald, had to sit in on the session and help with some of the recording, and one of the negotiated agreements was that I was *not* to have my name on the recording sheets or the tape boxes. That's EMI! Even today, the engineering on some of the Beatles' albums is credited to 'Abbey Road Studios.'"

THE MUSICIAN PRODUCERS

It's one thing to write a catchy song and a whole other thing to produce the whole package with its own definable character. These rare producer birds have developed entirely new

Susan Rogers on Prince's Process

"At the time I worked with him as his engineer, Prince did a lot of things that other people didn't do. However, he was most unique for his recording and production process. How many musicians did this, where one guy lays all the instruments and takes a song from beginning to end, with no preproduction or anything like that? He thinks of it, records it as he's thought of it, pretty much does everything himself, other than the horns. Oh, yes, and he would bring in Wendy and Lisa for female backing vocals. But pretty much do everything else himself. And we would mix it as we were going along. Print it and then the next day, after four hours of sleep, do another one. We didn't use automation, and we didn't use synchronization. At that time it was old school, 24-track maximum. His arrangement considerations were constricted by the limited number of tracks."

---

genres by writing and recording their own music. Now that's the way to shake things up! Nile Rodgers, Frank Zappa, Quincy Jones, Trevor Horn, Giorgio Moroder, Jeff Lynne, Lamont Dozier, Todd Rundgren, Robert Margouleff, Prince, L. A. Reid, and Babyface are some of the great songwriter innovators. And the younger generation includes Rodney Jerkins, Dan Auerbach, Linda Perry, and the Neptunes—hit makers with their own unique, recognizable styles.

LINDA PERRY ON CHANNELING MUSIC

"I approach production differently. For me, it's not engineering, it's not the tools and gadgets—it's right here in my heart. I am feeling the production and channeling the artist's music through me. The other day I had an artist in here to see if we could work together. I can't say yes to a project with someone I don't know, so I have to meet everybody. She was talking, and I looked at her and I thought, *Something just doesn't feel right about this girl.* I'm really sensitive about people. And I said, 'I get that you are a powerful girl, I see your strength. You've got it all together, but you are damaged. Tell me about the damaged part, because right now I'm not buying your story.' And this girl looked at me a bit shocked, and she said, 'OK,' and then she started telling me the real stuff. I thought, *There it is, this is the real shit.* And when she was talking, then I could hear music.

"I stopped her and said 'OK, let's go,' and I took her to the piano. I am always recording from the moment I sit down at the piano, and everyone sitting there always has a mic and headphones. I started playing and singing, 'Every day's a holiday in my head, *ba ba ba*'—and I'm just making up this melody and lyrics as we go, and I'm singing, 'I'm just waiting for you to join in.' Then I looked at the girl and said, 'OK, it's *your* turn.' And then she steps in, and it's a scary moment for her. Now she has to say what the fuck is on her mind! 'Oh, am I going to be brilliant? Am I going to say something clever?' But the point is: Who cares? Then, out of nowhere, she just starts singing the real stuff we were just talking about, and it just comes pouring out. And the song starts to form! She starts developing a melody as we are going along. I just keep playing it and playing it; then I stopped and said, 'OK, my job is done.' And I told her to go in the control room and listen back to everything we had just done and write it all down.

There is your story and there is your song. And literally two hours later, from when she walked in to when she walked out, she had a song and it's a great song!'

THE REBEL PRODUCERS

And then there are the "could-give-a-fuck" producers who insisted on taking their own path. I have the utmost respect for these true artists. People like Dr. Dre, Ross Robinson, Al Jourgensen—trailblazers to unexplored, dark lands. T-Bone Burnett insists on using ancient recording equipment for his undeniably recognizable sound. The Residents and Snakefinger created a genre of dark and humorous rock-type compositions including *The Commercial Album*, featuring forty one-minute songs. Steve Albini refuses to even be referred to as a producer, yet his recordings have his fingerprints all over them. Here are a few more notable "rebels":

CHRIS THOMAS

Chris Thomas enabled some fucked-up punks from the U.K. to take over the world, and even though their actual time on the stage was brief, it shook up a music industry bored silly with lame disco beats. Nothing like the massacre of Sid Vicious singing "My Way"! What a band, Sex Pistols. Changed it all. And that punk attitude is still going today. Chris Thomas, the enabler.

Figure 12-12. Kraftwerk's *Computer World* album.

Figure 12-13. Brian Eno's "Oblique Strategies."

KONRAD "CONNY" PLANK

Even though Conny wasn't credited for the production of the albums that fully embodied the Kraftwerk sound, Konrad Plank's Kling Klang Studio was where industrial synth pop was created, opening up the "Autobahn" to the world beyond Germany. I'm amazed at how fresh the *Computer World* album still sounds today!

BRIAN ENO

Strange, he is, and genius, creating a sonic landscape beyond pop music. But the rock monsters he created with Roxy Music, U2, and Talking Heads—all mysterious and otherworldly, due largely to this fellow. He introduced Talking Heads to African-inspired rhythms. He introduced minimalism to the musical landscape. And his production technique is revealed in his "Oblique Strategies" card set—a set of cards in a box with printed breakthrough ideas. I pull out my Oblique Strategies cards when I need to get out of a rut. Brilliant. Thank you, Brian!

MIKE PATTON

Mike Patton fronted Faith No More, whose 1992 album *Angel Dust* was considered the number one "Most Influential Album of All Time" by *Kerrang!* magazine in 2003. It spawned a whole generation of rap-rock artists, from Limp Bizkit to Papa Roach. Perhaps not the legacy he had hoped for, but Patton has gone far beyond what Faith No More achieved in the early '90s. Besides his pana-genre side projects, including Tomahawk, Mr. Bungle, Lovage, and Peeping Tom, Patton showcases his creative genius with his Fantômas project's album *The Director's Cut*. With vocal bloops, grunts, and eerie soaring notes, he re-creates the great movie themes using his voice as the main instrument. It is a clever and surprising ride. Beyond Fantômas, Patton has collaborated on several projects in the States and Europe, adding his voice and production sense to projects such as Björk's *Medúlla*.

Ross Robinson on Recording Deftones

"I was recording vocals with Chino from Deftones. We went in one of the side rooms in the very back of Sound City Studios. Everybody says it's haunted back in there, so I said, 'There's ghosts in here, Chino!' 'What? Really?' and he said, 'I'm getting out of here!' He was so scared, because the room had no lights, it was just dark and shitty. I said, 'OK, we're doing vocals in here.' And he's like, '*Whaaat*?' So I set the candles in, like, a satanic circle and he had to get in the middle of it to do his vocals. And the band was doing some kinda satanic séance out there in the studio. I put the mic on the floor and told Chino, 'All right, I want you to bring back the dead.' I don't know, I don't remember what exactly I said, but it was something crazy like that. I was in the room with Chino, and the song started and was rolling, but he wasn't singing yet, and I grabbed a candle because I was so excited at what was about to happen. And then I hocked the candle against the wall and it exploded! Wax went all across Chino's face, and he just let out a blood-curdling *screeeeaaaam*! Like, it was fucking *soooo* ripping. These were the three songs before Terry Date did their first record. Those three songs, in my mind, smoked what they did later. I tried to convince the band to rerecord that first record, but they weren't having it. It was a good first record anyway, but, um. Jeez."

Figure 12-14. Ross Robinson.

THE MAD SCIENTISTS

These are the producers who experiment with formula. They combine genres, choose repertoire carefully, select personnel, equipment, studios, replace band members for sometimes unknown reasons, make critical decisions that steer a music project in new and sometimes radical directions. I've been deeply influenced by these people: John Leckie, Adrian Sherwood, Butch Vig, George Massenburg, Flood, Nigel Godrich, Michael Beinhorn, Brendan O'Brien. Peter Collins created theater with Queensrÿche's *Operation: Mindcrime*. George Clinton's records always sound like a party. Trent Reznor's sample-laden industrial rock scared the crap out of me. And I loved it. And there is a newer generation of scientists in the lab: Danger Mouse, Timbaland, and Skrillex, to get started. But of all the scientists, this producer is still on top:

RICK RUBIN

This is the "fan" producer. He loves music. When he starts a project, he'll task a band to write three hundred songs so he can choose his favorite twenty-five to record. He loves creating musical controversy with fearless pairings of radically different genres. He'll handpick the environment and the personnel, putting everyone in a studio "petri dish" to see what develops. When I first began working with him, he had just finished the Geto Boys album, and he was proud of just how different it was. No one had rapped about killing and raping grandmas before—this was new territory. He was proud of that. And who else would have Johnny Cash paired with Tom Petty and the Heartbreakers singing a Soundgarden song? Well, that would be one of the greatest producers of our time.

Figure 12-15. Rick Rubin.

Marcella Araica on Timbaland's Beat Genius

"There was nothing more inspiring than watching Timbaland create beats from scratch. He would start by the tapping of a pencil or a pen on a music stand. Then he would tap on anything from a window panel, a mic stand—anything he could find that would mimic a sound in his head. It became such a phenomenon to me that we would challenge him by bringing in anything and everything to see if he could incorporate it into one of his beats, and he would. Water jugs, a rake, high-heel shoes, plants—anything at all! And he would be beatboxing the additional parts. A lot of the beats from Nelly Furtado's Loose album were created this way."

SONG STRUCTURE AND CONTENT

PREPRODUCTION

Preproduction can be both good and bad. It is generally a good idea to have mapped out several of the songs before going into the studio. It helps the session go smoothly and can save a lot of aggravation, fear, and wasted time, which can be very expensive. On the other hand, it can be a problem when the songs are so well laid out that they can be heard no other way. It is common that once you get into the studio and can actually hear the details in the timing, chord changes, kick drum patterns, etc., you will need to make changes. If the songs

Ross Hogarth on John Mellencamp's Control Room Writing

"I saw an interesting technique for creating song arrangements during the John Mellencamp records. John would always work up a new song in the control room, so that the first take behind the microphones could potentially be a record. Generally, the first play was usually just him and usually just G/C/D. Y'know, just *jang, jang, jang*—simple, no melodic hook, no real song structure yet. To me, his band at the time was as good as any of the greatest bands ever. It was Larry Crane and Mike Wanchic on guitars, Toby Myers on bass, Kenny Aronoff on drums, and John Cascella playing accordion or organ. There was incredible intuition within this band. So they would listen to John's song and Kenny would come up with the kick, snare, and hat pattern in the control room by tapping his foot for the kick, slapping his knee for the snare with his hand, and tapping his chest with his fingers for the hat. And he might play the toms on his thighs. Kenny would sit there in the control room, chart the song, and work out his thing. Then the whole band would sit together and work up all the arrangements, coming up with the hooks right there. This is basically how songs on the *Scarecrow* album were created. Everyone would come up with their parts instantly and instinctively, even though they hadn't played them yet. So when they went out into the studio to record the song, it exploded out of them! This technique ensures the recording won't be totally haphazard, but at the same time it leaves so much room for some kind of magic to happen."

Figure 12-16. John Mellencamp's *Scarecrow* album.

Geoff Emerick on Doing the Beatles, Song by Song

"With the Beatles' records, we worked on songs one at a time from beginning to end. All the way through mixing. And the sounds we recorded as we went along were the final sounds. Working this way would force you to make decisions. And with the Beatles it was just one loudspeaker in mono. It was really hard to get definition on two guitars coming out of one loudspeaker, giving them their own space so one guitar never clouded the other. It's easy when you can put one guitar left and one guitar right. Also, by doing it that way, when it came to the overdubs, if you realized an idea wasn't good, you would find a new idea immediately. If you tried the piano and then in the mix, if you realized the tonality of the piano wasn't right, we would switch to try the harpsichord. So you made decisions early. Since it was like the finished rhythm track, you could hear the gaps in the arrangement as you went along, and you filled them. And as you put it together, you were hearing the new, finished track. So there are no really superfluous overdubs on those things. All the parts are meaningful because they were thought up and applied at the time."

Figure 12-17. Kat Bjelland from Babes In Toyland.

have been so carefully worked out, it is difficult for the musicians to imagine the parts any differently. Another issue is over-rehearsal in preproduction. I feel it is best to leave the band a little unfinished and allow them to gel in the studio, finalizing the details and then recording. You need to allow for a bit of this polish. A little extra time will generally be needed to complete the outline of the song arrangements in the studio.

INTROS AND OUTROS

Often as a producer, I find that some of the big weak spots in musicians' writing are the intros and outros of their songs. These seem to be throwaway parts. Rock bands start on a riff and do that around eight boring times before the verse vocal comes in. Please be more creative!

George Drakoulias on the Black Crowes' Dodge Dart

"We were doing a song with the Black Crowes, and we crashed a Dodge Dart into a Dumpster. Over and over again. And we literally had a mic, an SM57, suspended inside the Dumpster. It was empty at the time. Ha ha! That was pretty fun. The crash made it on the record, too, on the start of 'Thick N' Thin.' You hear the engine, then the squealing of the brakes, and then the crash—."

Figure 12-18. George Drakoulias and Sylvia.

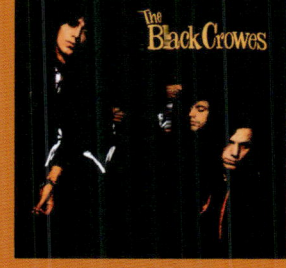

Figure 12-19. The Black Crowes' *Shake Your Money Maker* album.

Figure 12-20. The Girlfriend Experience writers obsessing over lyrics.

PRESSURE-COOKER HOMEWORK FOR THE BAND

Some of the most amazing songs are written *during* record projects, in the evenings, when the recording is done for the day. I enjoy tossing a pile of homework on the band to keep them occupied overnight. "Write a song for a film! Write a song for a video game!" Bands hate me for it at the time, and then love me for it later, when they have created a song they would have never written, had they been given a month to write it. Pressure makes diamonds, you know.

WRITE FOR ANOTHER ARTIST TO GET OUT OF A RUT

When you need your artist client to write a song that gets them away from territory they have already explored—i.e., all the songs sound the same, are in the same key, or the melodies all travel through the same set of chords or notes—then it's time to throw them a curve ball. I'll ask them to write a song for someone else entirely. Get, say, your metal band client to write a reggae song. Get your prog-rock band client to write a disco dance song with the same beat from beginning to end. Get your girl singer-songwriter to write a cock-rock song. Watch what happens—instead of the metal band bringing you a reggae song, they will bring you something that is still them, but with a new approach. They reach into different areas. Stretch their abilities. Even if the resulting song does not make the album, the exercise is guaranteed to get them out of their rut.

Jack Joseph Puig on the Black Crowes' Acid Session

"The Black Crowes all decided they were going to take acid one New Year's Eve. Well, I thought no way was I going to take the LSD, but I decided to make them all 'acid instruments' for the session. I spent two days prior to New Year's Eve making the craziest-sounding instruments I could think of, à la Tchad Blake in some ways. For instance, I ran a keyboard into a Leslie cabinet that had a snare drum in front of it that would rattle on certain frequencies. Then that would go into an octavider that would go to an amplifier that had a cymbal in front of it—miking it all in different ways—and on and on. So each instrument was something they had never heard before. Drums, vocals, guitar, everything. I called it 'the acid session.' I ended up using two bridges, an intro and an outro chorus from the session, cutting them into the masters on *Three Snakes and One Charm*."

Figure 12-21. Jack Joseph Puig

CLICK DECISIONS

The use of a click track can be challenge for both drummers and those who record them. Tool's music is full of tempo changes and slight fluctuations, nearly impossible to program into a click without completely ironing out all the feel of a performance. My solution was to document the tempos in the entire song, and then "play" a drum machine live during the recording. The drum machine, an Alesis SR-16, was programmed with a cowbell click that I would set with the first tempo of the song, and as a mapped tempo change was approaching, I would turn off the machine, changing tempo quickly, and then drop the machine back in on the downbeat of the next section with the new tempo. Some of the Tool songs had five or six changes, so I was kept real busy during tracking. If we missed a section, we would play that section's tempo on the click, the band would track to it, and I would cut that section into the master. (Two-inch tape, baby!)

Figure 12-22. Alesis SR-16 drum machine.

I still use the Alesis machine on occasion when there are multiple tempo changes in a song and I want to keep that live energy. I have yet to find a way that I can "play" the click out of Pro Tools so it hits on a downbeat. Sometimes on songs with multiple tempo changes and fluctuations, like on the progressive Americana Patchy Sanders project, I'll have the band play an entire song "freewheeling" (without click), and then I'll record each section separately using the Pro Tools click. After getting the desired performances for each section, I'll string them together using the transitions edited in from the freewheeling performance. Big tedious project, but well worth it.

On the other hand, I've been surprised by some drummers who do not use a click at all. Steve Ferrone is one such drummer. During the recording of Johnny Cash's *Unchained*, no one was playing to a click. In fact, it was just Johnny with Tom Petty, Mike Campbell, Benmont Tench, and Howie Epstein sitting around in a circle, playing songs. The drums were put on later. Steve Ferrone listened back to the multitrack and played along to it. His performance tied it all together. Amazing. It sounds like everyone was playing to a click. That is why those cats are *professionals*.

Figure 12-23. Skunk Anansie's drummer has a great sense of humor.

Bob Ezrin on Drummers Playing with Themselves

"If the drummers enjoy it, sometimes playing with a click frees them up to simply concentrate on the music. The trick sometimes is to use something other than a straight click. We often get lazy and just give drummers a stock *tick tick tick*. But the best 'clicks' are sometimes the ones they make themselves. I would often get a few bars of straight time from the drummer, or some combination of their stuff—sometimes just their hi-hat. And I would play that to them as a click. Because it was their own sound, they would tend to lock into it very naturally. Don't take this the wrong way, but there's something about playing with oneself that just can't be beat!"

Ross Robinson on Using Delay Instead of Click

"I don't like to use click, because I do a 'click track' differently. I will set the song's tempo on a delay and then put that delay on the snare drum or on the vocal mic just humming through the headphones as they're playing. That delay gets in there and starts swinging like crazy. It can be with the most insane music, and it just goes *whooooof!* And they love playing to it. It's so fun, and it works like a charm! You can speed up or slow down, and the delay will find its way back in. You're never catching up to some strange alien machine."

Figure 12-24. Don Henley.

Ross Hogarth on Don Henley's Psychology Lesson

"So we get Don Henley's session going up at his house, and Stan Lynch (Don's producer) says to me, 'If I can get Don to play drums on this song, then we'll be able to keep this track. If I don't get Don to play on this song, there will be no DNA of his on the track, and it probably will not move forward.' Stan continues, 'So, I'm going to start by sitting down at the kit myself and play drums, and I'll just continually get worse and worse, just to piss Don off.' This is how well Lynch knows Henley. Stan says, 'By the time Don gets so pissed off with me, he'll either leave and bail, or kick me off the stool and then we'll get him at the drum kit.' Ha-ha! Stan obviously knows his artist really well, because Henley is really difficult, so we get the drum sound, Lynch sits down and starts playing on the song, and it sounds like Tom Petty! It just does, the way he hits the hi-hat, the feel of the rhythm, the military grip, the whole thing. (Stan Lynch was the original drummer in Tom Petty and the Heartbreakers.)

"And sure enough, by about three or four takes in, Henley is screaming at him that he sucks, and I can't see Lynch because he's way down the hallway on the drum kit, but I knew Stan was smiling, because this is what he wanted! He knew that if he could get Don pissed off enough, he would kick him out of the stool! So Don is having me turn the drums down on the control room monitors, and these are great players on this session. Bob Glaub on bass, Stuart Smith on guitar—I mean, really good players wondering what is going on.

"So finally, Henley goes down into the main room where the drums are and kicks Lynch outta there. Don is literally, at that point, so frustrated with Lynch that he just says, 'I'm outta here!' and jumps in his car and takes off. Lynch goes, 'We got him!' I asked, 'How do you know we got him? He bailed!' Lynch says, 'No, no, no. We got him!' Literally seconds later, Lynch's phone rings and it's Henley. He's now on PCH, but he's calling Lynch, and he says, 'Order some food, I'm coming back.' So, he comes back to the studio, we order food, he sits down at the kit, and if you ever watch Don Henley play, he's real 'upright' and those tom fills are so distinctive. You know what I mean? Totally not like Stan, the way Stan holds his sticks or the way Stan plays. Now, I haven't changed a thing about the drum kit or recording setup, but all of a sudden it sounds like a Don Henley song. And the track was saved."

Matt Wallace on Being an Idiot with Faith No More

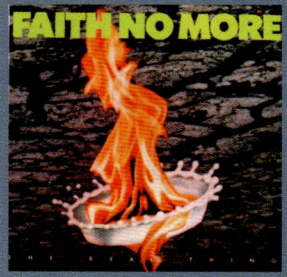

Figure 12-25. Faith No More's *The Real Thing* album.

"Sometimes as a producer, you just have to look like a total idiot so the artist will forget to be nervous. On all the early Faith No More records, we didn't have budgets. When they were signed to Slash Records and we were suddenly able to record at Studio D—a really nice studio in Sausalito, California—it kind of brought the pressure up. The album was *The Real Thing*, and we were trying to do the song 'Epic.' Mike Bordin, the drummer, would sometimes start thinking too much about his parts, and it affected his performance. So I went out there in the studio, and I took my shirt off and started doing some stupid stuff, basically doing a nerdy striptease, to distract him! And for a minute he forgot he was in the studio. It took the pressure off. He was just looking at me, saying, 'You idiot, you're dancing around taking your clothes off. I don't know why you are doing that!' It was just enough silliness that he forgot to think about drumming and he just did it."

HOOKS, HOOKS, HOOKS

I love challenging a band to write new and better hooks. If the intention is to have a pop hit, then a solid lyrical hook should be a requirement. A simple and memorable three- to seven-word phrase that repeats throughout the song—this is what I will usually ask for. The idea is that the listener can hear the song once and remember the hook lyric. If you have that, then you know you have a winner. Pop lyrics don't need to be profound. The pop hook lyric doesn't even need to make sense! One of my favorite examples of a pop song loaded with hooks is the Offspring's "Pretty Fly (For a White Guy)." Hooks include girls singing "Give it to me, baby!" a yelling hook that goes "A-ha, a-ha!" and my favorite low-voice "Uno, dos, tres, cuatro, cinco, cinco, seis"—frickin' brilliant. Imagine the song hook to be like Arnold's memorable *Terminator* catch phrase, "I'll be back," or Clint Eastwood's "Go ahead, make my day." And why do the hooks have to be lyrical? They can be melodic vocal ooohs and ahhhs. They can be "hey, hey, hey!" yells. They can be repetitive melodic guitar or key lines or even honks, dog barks, bells, or snaps. Just repeat the sound or sonically frame it in a way such that the listener focuses on that part only. Hook 'em!

LYRICAL FODDER

The best reference material to kick-start the lyric writing process? Here are a few must-haves: slang dictionary, thesaurus, rhyming dictionary. OK, sure, but what about other inspirational tools? Like cookbooks, sports magazines, or technical manuals? These could inspire lyrical hooks like: "recipe for a broken heart" or "on the beach but barely there" or "I'm not your motherboard" or "everybody knows everything and nothing." Pretty much anything can be used for lyrical fodder. Sit in a restaurant and listen to the chatter in the next booth. Listen to the waiter taking orders. Listen to the fast-order cooks rattling through their routine. I hear lyrics already: 'You're always better early in the morning" or "Just one taste of paradise" or "I won't stay hot forever." So maybe they all sound like country songs, but you know what I mean!

Figure 12-26. Brad Wood.

Brad Wood on Writing a Bridge

"Do you have a bridge? Your middle eight is the most important part of the song! It's your chance to put up a different perspective. If you have lyrics about a particular thing—say you are mad at someone—how about a change of view for the bridge? Some self-doubt? Or how about someone else's point of view? You have eight bars to do something really amazing here. Why just grind through a solo?"

BLENDING THE PALETTE

Each country has their version of musical instruments that are just familiar enough that you could probably make a sound out of them. So why not try? Or bring in a player with a set of vibes, a hurdy-gurdy, or a lute. Add unusual unidentifiable sounds into your production. Doing so will give your recordings a completely unique character. The world is full of incredible sonic color—even if the tracks don't make their way into the final mix.

Figure 12-27. Emilio Solla's band.

Emilio Solla on Blending Musical Flavors

"Our lineup is quite unconventional, with one foot on the folk side of Argentine and Latin American music, by means of the accordion and violin; on the other side, you have a jazz trio plus. There is a rhythm section with double bass, drums, piano, and then you have a full horn section of four guys—trumpet, trombone, two saxophones—a bass clarinet, flutes, and stuff. I didn't copy the lineup from any other band I know. I just thought of which colors I wanted, and I made up the mix of instruments. So I think the flavor is quite atypical, a new kind of sound. I would say it is 'South American jazz.' Modern tango meets jazz. Jazz with a very strong Argentine thing."

Hans Zimmer on Roma Music for *Sherlock Holmes*

"In the first *Sherlock* movie, I used all these gypsy violins or 'Roma' violins, as they are called. And on the second movie, I was thinking that I basically know nothing about the Roma culture where the music came from, so I said to director Guy Ritchie, 'I want to go on this road trip to Eastern Europe and find out what this music is about,' and we went. I felt I needed to know them. So every day we were up at five in the morning to get into a bus and go to the next Roma settlement. Well, you would not believe you are in the middle of Europe or even in the Western world when you have seen such poverty and such discrimination—it's just horrible how they treat these people, and the discrimination is celebrated. And there were these just amazing musicians everywhere. This is the music of their great-grandparents and their great-great-grandparents. When we got to sit down and play music with them, all those reasons that I believed that music was the universal language—well, they all came true. We would write them a theme and then go out into a room with them and jam on it! The clarinet is basically the shredding guitar of the Roma music. He never stops!"

Ross Robinson on Classical/Punk Genre Bending

"On the album for My Own Private Alaska, it was this classical piano with punk hardcore drums and a singer. Their theme was 'three guys sitting,' and it was the most miserable music ever. So dark. I was wore out after that one. And they're French. So they're like *sooo* expressive—'I KIIILLLLL.' Oh, God, constantly killing chicks over love in their songs! It was gnarly."

THE IMPORTANCE OF "NEGATIVE SPACE"

Here is an important production tip: As you create your sonic landscape, take a moment to listen for open space. "Space" in an artist's canvas is as important as content. Before you call your recording finished, listen for something you can take away to allow the track to breathe more. Space allows for dynamics in music. Space allows the listener to notice the background, to notice the air around them, to notice their own breathing. Take a listen to John Cage's "4'33"." It will put a smile on your face as you realize that he has just made you part of the music. And you're on the receiving end!

GET THE DAMN RECORD FINISHED, SET LIMITS

Perhaps the biggest challenge for anyone recording music today is just getting the project finished. Not posting it for anyone to hear until it is ready, and then knowing when it is ready. By putting a time limit on a project, you are creating an ending. With pressure to end the project, you are forced into being creative. This is a time-tested technique used by artists and musicians for centuries.

Eric Valentine on Going All the Way

"I've designed my process so there are no real limitations on creative development. I do this by producing with all-in funds, so it's a flat fee at the beginning of a project and I just hand you a finished record at the end. Nobody's thinking about accounting or budgets, so money pressure does not get in the way. This is not real great for finishing projects on time, but really great for the creative process. If somebody has an idea that might make the song better, I want to try it, so every idea gets tried! Now, with computer recording, it is possible to try everything, because there's room for everything. That may be causing problems for me, at this point! These days, at the 80 percent point on a project I've usually overdone everything, and I start scaling the recording back. But I like having those options of going all the way creatively!"

THE GOOD, THE BAD, THE UGLY

Face it: Some days you will get nothing done. You will spin your wheels all day. The singer will have a cold, the drummer will be missing, the computer just doesn't boot up properly—you are just gonna have those days. Accept it and recognize early that if it ain't happenin', you should just take the day off. Better to leave the studio and do something else productive or fun, rather than beating your head against the wall when the universe is working against you. Things to do when it ain't happening in the studio? Oh, to hell with it. I'll go watch TV. I'll go shopping. I'll make a nice salad. I'll go for a walk in the park.

Figure 12-28. Jefferson Airplane.

Al Schmitt on the Airplane, Hot Tuna, and Getting Dosed

"When I was doing the Jefferson Airplane, they had one guy that all he'd do was roll joints. They had a nitrous oxide tank in the studio with four hoses coming out of it, so four people would be getting high on it at once. And this was a regular thing. And they would dose everybody. You had to be really careful. The way they got their manager was, they injected LSD into a Coke can with a hypodermic needle—so they loaded it without even opening it. So he opened it and drank it! One time, during the wedding of Tom Donahue, who was a very famous disc jockey in San Francisco, they spiked the wedding cake! Tom got married at the Jefferson Airplane's house—2400 Fulton Street in San Francisco, right across from the park. I was there with Tommy LiPuma, and I had to convince him—'Don't touch the cake.' They got me on the Hot Tuna record while we recorded the acoustic show up at the New Orleans house in Berkeley. Outside of the club was Wally Heider's remote truck and I was getting my papers together, ready to take notes for the start of the show, drinking an apple juice. And suddenly everything started to go crazy psychedelic in my vision, and I knew they'd got me. I turned to my engineer said, 'You're on your own.' And I just sat back and listened. From then on, if I ever took my eyes off a drink around them, I'd throw it away."

BREAKTHROUGH PRODUCTION IDEAS

Song just laying there? Time to pull out the sound effects library. Just scrolling through a list of sounds online could be enough to kick-start a sluggish day. How about a church bell? How about a backward church bell? Or record something else backwards. That's right. It doesn't even have to work. Just do it. Reignite your fascination! Strip all the drums off a mix and rebuild the rhythms using a slamming car door and trunk instead. Trunk for the kick, car door for the snare. Clink a glass of water for the hi-hat. You don't even need a drummer to do that! And when you are done, go ahead and add the real drums back in—could be very interesting!

Figure 12-29. Sylvia's first album: The Beatles' *Revolver*.

Geoff Emerick on Lennon Discovering "Backwards"

"John had a reel-to-reel tape recorder—of course there were no cassettes around the time of the Beatles' *Revolver*—and the only way you could listen to what you'd done that day would be with a playback reference reel. So anyway, John took a copy of, I think it may have been the track 'Rain.' He took it home and came in the next day and said, 'There's something wrong with the tape. It doesn't play properly.' Well, John being John had threaded it up backwards, so he was listening to it backwards not realizing it. Once he found out what he'd done, he wanted to try everything backwards. He wanted to try guitars backwards, try vocals backwards, etc. Just for fun, I even suggested that we get someone speaking Russian language and play that backwards to see if it was English forwards. So this was a new thing. No one had ever had anything running backwards. So that's how the backwards guitar thing started."

Figure 12-30. XTC's *Black Sea* album.

Nick Launay on XTC's "Respectable Street"

"This is a true story, even though it might sound like I'm making it up. I was the assistant on the *Black Sea* album by XTC. Terry Chambers, who was the drummer of the band, was a quiet fellow. He didn't say much. But occasionally he would come up with an idea, and they were all usually very funny. And he was always very insistent about it. There is this song called 'Respectable Street' on *Black Sea*. At one point in the song, the singer, Andy Partridge, sings, 'Heard the neighbor slam his car door / Don't he realize, this is Respectable Street?' So Terry says, ''ere, so I got this idear, right? Why don't we put microphones out in the street, an' record a car door slammin'?' Which is kind of like, so literal, right? But, OK, do we have to put the microphones out in the street? 'Yes, we do!' Well, I thought, why don't we just drive the car up to the nearest door of the recording studio and slam the door there? Right? No! Not good enough! So at four o'clock one morning, there's me, with God knows how many microphone cables, one after the other—I think I used something like fifteen of these fucking things, because Terry wanted a microphone on the other side of the street! So we literally had one mic near the door and the other mic on the other side of the street. Luckily, there was not that much traffic. But then we also had to have headphones for him, because he had to do it himself with the song playing so he could slam the door in time! So it was this whole elaborate thing, completely unnecessary. He could have gone out there with a tape recorder, recorded a door slamming, done, and then flown it in—but no! It was hilarious, it was fantastic, and here I am talking about it because it's one of the most memorable things I've done."

Geoff Emerick on the "Yellow Submarine" Brass Band

"The studio used to keep these stock brass band recordings in the sound effects library. So we took a Souza march and copied it and chopped the tape up into little bits and threw them up in the air; and when they landed, we picked them all up and stuck them together again! Backwards and forwards, whatever way they ended up. And when we played it, I had put them back almost into the same order that they were to begin with! I actually had to fade out the last note because of the copyright. There's one little reverse bit in the middle where you hear it go backwards and disappear. That was on the Beatles song 'Yellow Submarine,' right before a chorus. We did a similar thing on 'Mr. Kite' with a brass band, and we also fooled around with double-speed calliopes and xylophones and stuff like that from the sound effects library."

JUMPING THE SHARK

There was the TV show *Happy Days*, with the coolest of cool characters, the Fonz. The show was at the top of its game. Then one day, they aired an episode in which a water-skiing Fonzie did an acrobatic jump over a super fake-looking shark. Like that was cool? Well, everyone knew at that moment that it was all over for *Happy Days*. The TV show was no longer cool after the Fonz "jumped the shark." Viewership declined, and the show was finally pulled off the air.

In the world of production, there will come a moment when your production will have jumped the shark. It is usually around the time you bring in the children's choir. Might be a challenge to bring you back after the overindulgent fake plastic shark appears.

Bob Ezrin on "Overproduction"

"The problem with the current set of tools is that it's very easy and fast to record and edit now, so we tend to do as much of it as possible and then wait to make decisions later. But the thing is that as we are adding to the tracks, we seldom mute previous stuff, so we just get used to layers and layers of additional information, until the song's natural state involves hundreds of tracks. I'm as guilty of that as anyone. But I try to discipline myself to go back and do some culling before getting into serious mixing. I think often that in the analog era we were forced by the limitations of our medium to be more creative than today. We not only had to figure out how to attain the sounds for which we were searching, but also how to shoehorn them into the limited space we had for recording and storing the material. We had to make choices on the fly. If we wanted A, we often had to sacrifice B to get it. I did enjoy the 'Chinese puzzle' element of creating recordings a few decades ago."

GREAT MIXERS OF THE WORLD

 ANDY WALLACE
 JACK JOSEPH PUIG
 CHRIS LORD ALGE
 CHUCK AINLAY
 DAVE PENSADO
 MANNY MARROQUIN
 MARCELLA ARAICA
 JAYCEN JOSHUA
 BOB CLEARMOUNTAIN
 JIMMY DOUGLASS
 ALAN MEYERSON
 MICK GUZAUSKI
 PHIL TAN
 JIM SCOTT
 ANDREW SCHEPS
 DAVE REITZAS

Figure 13-1. Great mixers of the world.

13

MIXING

PERFECTION MIXING

After recording a big mess of stuff, there is always the tendency to want to hear every little bit of it. I'm one who thinks it's not so important. Let some things fall away into the back-

Figure 13-2. Dave Pensado rules.

Dave Pensado on Changing Problems Into Opportunities

"I did a mix for Linda Perry, at a time when she was kind of 'reinventing' herself. I listened to the tracks and realized there was a distortion problem on parts of the vocal. I had to decide: Do I tell the client about this problem or find another solution? The best way I thought to get around this problem was to add distortion on all of it. When distortion comes and goes, it catches your ear. But if the distortion is always there, no one can tell the difference! I had read that Dave Way had used a SansAmp in a mix, the little black foot pedal with the little dip switches. It was one of my favorite guitar pedals at the time, so I judiciously recorded the rest of the vocals through the SansAmp, and I probably get more compliments on that vocal sound than on just about anything else that I've done!"

ground. Is it so imperative to hear every hit of a hi-hat? Probably not. Just because you know it is there, doesn't mean it has to be heard. Or even be in the mix at all. Dense mixes lose dynamics, and stripping away tracks during sections of a song can be a painful exercise, but may dramatically help. Be brutal. Go ahead and tear that recording apart! Cut and mute and redesign! Give the song what it needs and not much more. If it needs a lot, then it needs a lot.

REPAIR MIXING

With luck, if you tracked it from the beginning, you won't have a lot of fixing to deal with. But "fixit mixing" is unfortunately the norm. Trying to sum a dozen of someone else's out-of-phase drum and guitar tracks in every song is a bummer and eats precious time. Ugh. Of course there is no "right" or "wrong" way of doing things, but there is the "suck" and "not suck" way. If you are presented with ten tracks of the same guitar performance—same player, just different microphones—turn off all but one of those mics. If the engineer is too lazy or does not have the equipment to sum when recording, those redundant mics get forfeited. Same goes for drum mics and whatever. Just strip it back to one mic per performance. Start there. Another thing to do is run the whole thing into a compressor and smash the hell out of it as your first step. That may shave hours off your whole mix. Do the same thing with a stereo EQ: Put the whole mix through it and do a serious A/B comparison with a recording you know and like, maybe a commercial track from a popular artist, or a vintage track from a legacy artist. There—you're already most of the way to the finish line. Oh, except now here comes the committee.

Figure 13-3. Wade Steigmeyer and Fornax Chemica getting deep into mixing.

MIXING BY COMMITTEE

On a "democratic" mix session, what does a guitar player want? More guitar! What does the drummer want? More drums! Mixers struggle to make everyone happy and wind up with extreme "perfection mixing." No dynamics, no background. Everything right up front. The only thing worse than this type of mixing is a spiderweb in your face. Ick! Some bands think decisions should be made "democratically" like they're attending a PTA meeting—"what's best for our child." Was it the Beatles who started this? Where everyone got a song on the record, even if they weren't great songs? Someday I'll take only gigs where I make all the mix decisions—ha-ha! Watch out!

ADVENTURE MIXING

If that day comes, my mixes will be out of control. I'll connect every box, process the whole thing through a spring reverb, have shit spinning around in crazy panners and multiple delays. Hell, yeah! Those mixes will be unlistenable! Ha-ha! It happened on an Ingrid Chavez track I mixed

Shelly Yakus on Mix Dynamics

"A song production needs to have the dynamics of a good suspense movie. *The Sixth Sense* is a perfect model. You don't want to start the movie on '10'—that would make it boring later on. You want to build it with plot dynamics. Give it tension and release, and give it a climax. In a song, you'll want to build the verses and choruses just like a movie plot, until the payoff in the bridge. So as a mixer, even though you're not the producer, you need to understand how to take the producer's creation and get the most out of it, so the listener gets the message."

Figure 13-4. Rich Veltrop mixing in Studio B at RadioStar.

Matt Wallace on Thinning Out a Mix to Make It Bigger

"The problem with Pro Tools is that everyone is postponing commitment. Now people have a hundred twenty-eight tracks when it is time to mix. If you've ever worked with Chris Lord-Alge, he just does his thing, and you might say, 'What about that guitar-zither thing?' And he'll say, 'Well, I didn't like it.' And you'd say, 'But we spent two days on it!' He's like, 'I know, but it doesn't work in the song.' And sometimes you need that person to tell you, 'I know you guys spent three days and $600 on recording this and that, but honestly, it doesn't help the song.' And it will get left out of the mix. And people get upset! I have to tell people, 'Listen, if you want the patient to live, you have to cut off a foot. I know you'll miss it, but otherwise the gangrene is going to spread and the patient will die—so, cut it off.' And people don't like it! They just put too much stuff in the mix and want to hear every little thing. Then you have a cluttered mix and everything sounds small. Remember, the less stuff in there, the bigger it sounds."

George Drakoulias on the Bangles and Panunzio

"We were mixing with engineer Thom Panunzio—he was Jimmy Iovine's guy, a real hot-rodder.... We were mixing something for the Bangles, for a soundtrack. It was 'Hazy Shade of Winter,' a cover of a Simon & Garfunkel song. On the end of the song, the vocalists sing, 'There's a patch of snow on the ground,' and we said, 'Hey, what if everything keeps going after the vocals stop and everything just gets airy and weird?' This is the type of thing you decide to do at four in the morning, and Thom's eating an apple, and he says, 'Oh yeah, I can do that.' I ask, 'How you going to do it?' and he says, 'Don't worry. I'm going to do it.' So we are running the 24-track master, printing the mix, and here comes the part. 'There's a patch of snow of the ground'—and he had the apple in one hand and he took his finger, and he pulled the multitrack right off the capstan while the tape was rolling! We were all freaked. 'Are you crazy?!' But it worked! He pulled the master tape off the playback head just enough that it made the track go weird, and the echoes and everything else were still going. And it didn't sound like he had just stopped the machine! That was one of the greatest things I've ever seen. Can you imagine? You've worked on this thing all day, all night, and it's four in the morning, and this guy does this and you're like, 'He's gonna break that!' Too bad the band didn't like it, but we liked it."

Brad Wood on Mixing

"By the time I cut vocals on a record I've produced, it's already really sounding close to mixed. I'm monitoring through what will probably be my mix chain from the start. And I mix stems into some analog compressors, and those are always engaged, always from the start. Even if I'm using some L3 on the final mix, it's on there because I want to hear it as well. I don't fuck around. I don't have time for that. I have to be able to move quickly."

Bob Clearmountain on Long-Distance Mixing

"I can mix an album at home while the band members are also at their own homes listening to the mix being live-streamed while they type comments to me over iMessage or Skype. The guys in Barenaked Ladies said it was like typing comments into their 'Clearmountain Mixing Software.' Bryan Adams, while on tour somewhere in Europe, mixed a song with me wearing headphones plugged into an Apogee One connected to his MacBook Pro while sitting in a Starbucks. With Bruce Springsteen, we still use our DolbyFax ISDN boxes, which only have a maximum latency of 150 ms, so we can see each other on video iMessage and he can, if need be, actually give me a visual cue when to turn something up or down. It's really like he's sitting right next to me, but he's in New Jersey and I'm in L.A.! It's bidirectional, so we can speak to each other with our talkback buttons."

Shelly Yakus on the Digital Mixing Street Fight

"There is a lot to like about digital recording. What I don't like is that you can't use two hands to mix. You've got a mouse with one hand. Who the hell mixes a record with one hand? You really need two hands to adjust a compressor; you can't do that in the box. It's like having a hand tied behind your back and being thrown out into a street fight. What are you going to do with that? Every mix is a street fight!"

Larry Crane on Accidentally Mixing the Go-Betweens

"I have this Alesis MidiVerb that I crave sometimes. I think to myself, 'Gee, I want preset no. 4.' Once I accidentally did a bunch of final mixes for the Australian band the Go-Betweens, on their album *The Friends of Rachel Worth*. I had done all the rough mixes, and they went to Germany to do the final mixes, but ultimately liked my rough mixes better. They wound up releasing the roughs, and that Alesis MidiVerb was the only reverb used on those mixes. Sometimes mixers just try too hard, plugging in a bunch of cables, running everything through an EQ, and boosting 10K. It just starts to ruin it. Instead of being 'smart,' they should just let it be what it is. Skip the 'dog and pony show.'"

Figure 13-5. Larry Crane of Tape Op and Jackpot Studios.

Visual Mixing Concepts

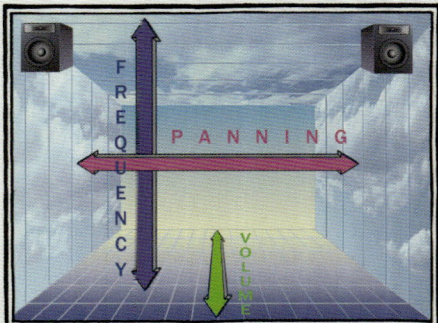
Visual Concept

His ideas are original and revolutionary. If you can imagine what a mix "looks like," David Gibson has found a way to describe it through illustrations. David is the founder of the Globe Institute of Recording and Production, with courses on these visual mixing concepts. Each image describes the depth, width, and elevation of individual instruments within a mix for a particular genre of music. Genius!

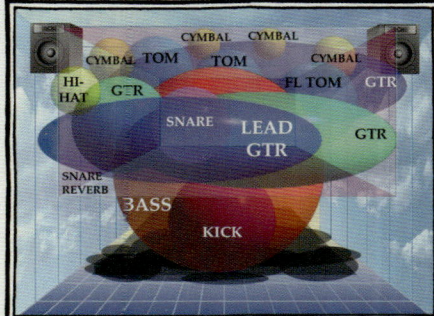
Alternative Rock

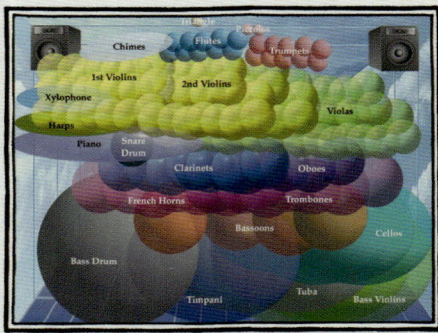
Classical

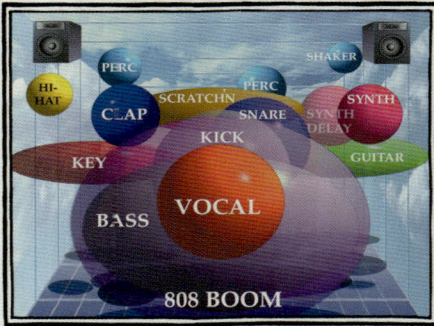
Rap

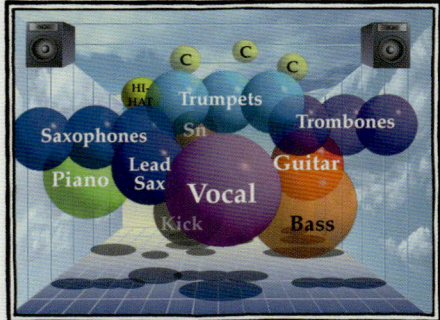
Big Band

Figure 13-6. David Gibson's visual mixing concepts.

Eric Valentine on Modern Mixing

"Because of computers, anything in a mix can be changed or recalled at any time, at any point in the process. That's just expected now. There are record company execs that insist, 'I will not hire anybody who mixes on a console because we have to be able to recall stuff. I only hire people that mix in the computer.' That distinction has been made at this point. I love analog, but I have to be competitive in this way also, so I find a way to balance it out."

for Prince. He said, "Mix it!" and left the room. I sat there wondering if I should play it safe and do a nice "perfection mix." Instead, I said fuck it and plugged in a dozen crazy delays and had shit flying back and forth, loud as hell in the mix. He came in and listened. I was so nervous, I was twitching. He nodded his head. I think he liked it a lot. Soon after that, I was offered a job at Paisley Park. I didn't take the gig, and he didn't use my nutty Ingrid Chavez mix. C'est la vie.

BIG DUMB TOM PANNING

During the Tool records, I panned the toms of Danny's drum kit across the L/R audio field, small tom to big tom. I might not do that today. It just seems so mundane and predictable. Better would be to narrow the field, or heck, mono the toms. When you stand in front of a drum kit and close your eyes as it is being played, you are not hearing the toms flying across the room left to right! Sometimes the perspective of the drummer sitting inside a kit surrounded by the toms is appealing, but I'd rather have the drum instrument in one position on a stage, with other instruments surrounding it.

LEFT, RIGHT, CENTER

Direct center, hard left, or hard right can make for better instrument and vocal positioning in a mix. Try the dead center, or hard pan your mix elements and layer tracks directly on top of each other, differentiating each of them through EQ, not placement. Whenever I can, I'll

Figure 13-7. Jim Messina.

Jim Messina on the Doors' Dead Rat

"Sunset Sound Studios had some legendary echo chambers. I would be in the shop while Bruce Botnick and the Doors were recording next door, and I would hear, 'People are strange / When you're a stranger' coming out of the chamber, over and over and over again. It was driving me crazy! After a while, Bruce poked his head into the shop. 'I keep hearing some weird noise coming through the chamber returns. Can you investigate? It's screwing us up.' So while he and the band took a break to go to dinner, I went into the echo chamber and, lo and behold, there was a dead rat lying there in a puddle of water. Apparently it had gotten trapped, and the sheer volume of the 'People are strange' noise being pumped into the chamber had driven it mad, and it just keeled over. I picked it up by the tail and thought, *What am I going to do with this?* And I don't know why I did it—maybe because they we driving us all mad—but I just laid it across the keyboard of the band's tack piano and left it there in the session. I figured it would make for a good laugh. A little while later, Bruce came in: 'We're leaving!' I asked, 'Aren't you going to work anymore tonight?' He said, 'No, Ray's on a major, major bummer. He came in after dinner and there was a dead rat on his keyboard. He took that as a bad omen!' Oh God—I never intended that to happen! I just kept my mouth shut and never told anyone that it was me. I never gave it up."

Figure 13-8. Studio Technology AN-2.

commit to one of these three positions. It is a trick Rick Rubin taught me. Maybe it came about because the old analog consoles did not have variable panning pots, only L-C-R as a choice in the monitor section. It just sounds strong and confident this way. The L-C-R panning in the mix also gives the lead vocals an important opening in the sonic panorama. Big dry vocals, right in the middle. Right in your face. That is a Rick Rubin thing.

I use an external rack-mount Studio Technology AN-2 to widen the vocals in mixing. It is a simple box with just a few settings. I love the sound of vocals big and dry and right smack in the middle of a mix picture, and the AN-2 helps to create that effect.

Recording tip: When initially recording a drum kit, try balancing the toms by monitoring the returns in a mono position when checking tom mic levels. This is the best way to listen for consistency and phasing. Then, no matter how you pan the toms later in your mix, they are always perfect.

RE-AMPING WITH THE "DOUBLE-DI" TECHNIQUE

Figure 13-9. John Cuniberti's "re-amp" box.

Here is how you can take a signal from a mix and reprocess it through a guitar amp. You might want to do this to add excitement to "boring" tracks that were recorded direct. To do this, you'll need to mic up a guitar amp in the recording room and send the targeted "boring" track through a "double DI" out to the amp, recording it on a separate track and blending it into the mix.

Bass guitars are frequently recorded direct to a track without an amp. In the mix stage, this is your chance to give the bass the excitement of having been played through a physical amp. In the case of re-amping a direct bass track, I would probably want to use a classic Ampeg SVT bass rig for the re-amp, but using a Marshall guitar head might also be an excellent choice for a re-amped bass.

The "double DI" will drop the impedance of the "boring" signal, then bring it back up as you send it out to the amp. To do this, you'll need two DIs. Take the "boring" signal you want

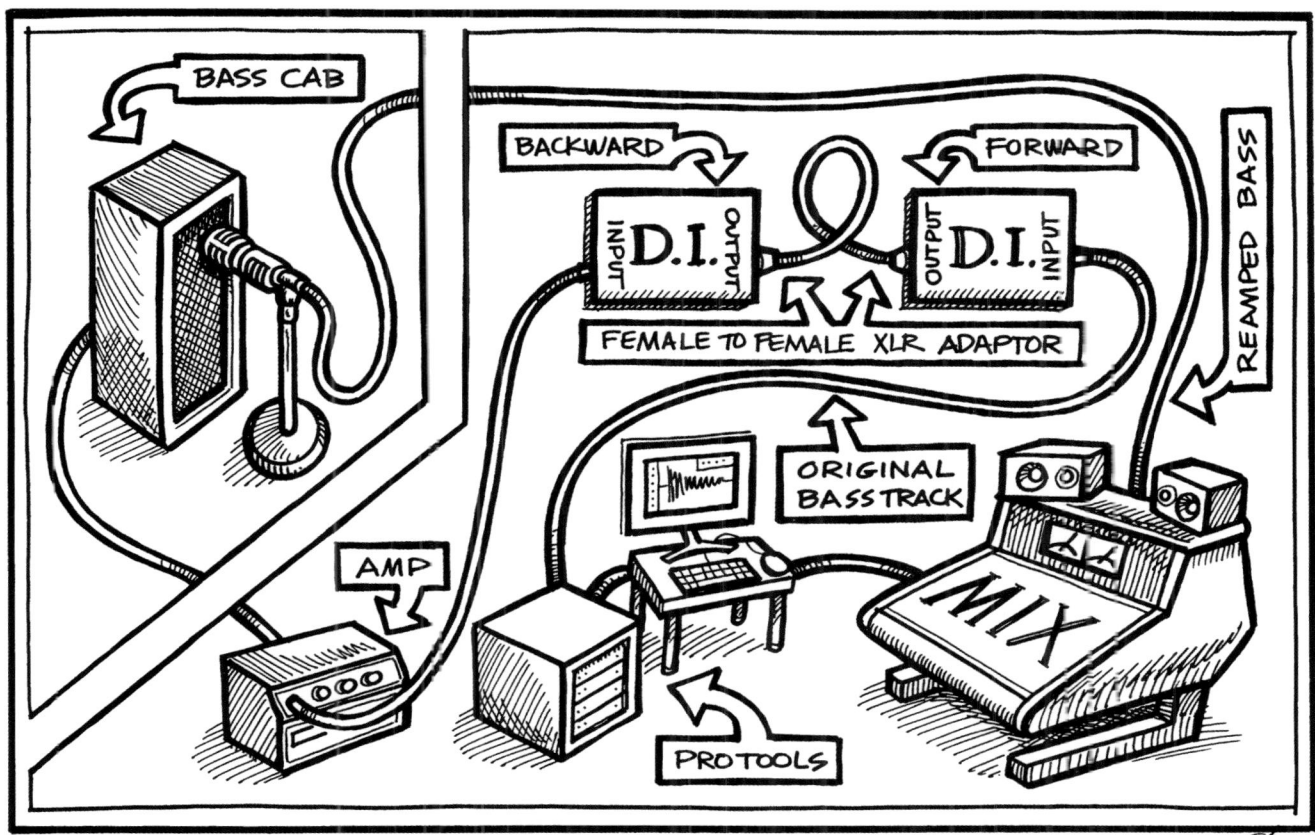

Figure 13-10. Re-amping a bass guitar signal.

to re-amp from the recorder and patch it into the "instrument input" of DI 1. Then, from the "mic output" of DI 1, use a mic cable with a turnaround and patch it into the mic connector of DI 2 (technically the "mic output"). Then patch out of DI 2's "instrument" jack into the amp. And there ya go! The "double DI" allows you to send a long cable without loss.

John Cuniberti's Radial and Jonathan Little's PCP are "re-amp" boxes to make this process easy and one-step.

MATT WALLACE ON RE-AMP CHAMBER REVERBS

"A great way to get a 'nonlinear' type of room reverb is to build one. When I first started out, I would take a send from my Carvin board and run it into my bathroom. I had a bathroom that had a stairwell above it, so the walls had angles to them. I painted the whole bathroom with gloss enamel, so it was very, very live. Then I put a speaker in there, sending the sound from the Carvin into the speaker. Then I'd put a mic in the room to get the reverb back to the mix. This is a simple way I was getting a really cool chamber-room sound. Then one day I thought, *Well, let's just try something else.* So I took the speaker and laid it on its side, took a snare drum and put it on top of the speaker, and I duct-taped it on there! I put a mic above it and recorded it! When you send a snare signal out of the board into the speaker, it sounds like another snare drum is being hit in the room. You can hear the snares rattling. Sometimes you can flip the snare

Figure 13-11. Re-amped snare.

Ross Hogarth on Re-Amping Keb' Mo'

"I find a lot of acoustic instruments really benefit from re-amp distortion. I call it 'happy' distortion. Dobros, acoustic guitars, and harmonicas especially. Also drums, percussion, shakers—even vocals. However, anything that's already got distortion doesn't re-amp real well—like re-amping an electric guitar into another electric guitar isn't very cool. But anything you can put happy distortion on and make stand up uniquely in a track really helps. On the Keb' Mo' record, we had a kind of a straight-sounding harmonica. It was just recorded cleanly onto the track. Now, if you listen to Little Walter, the sound of his harmonica is not the clean Willie Nelson–style sound—it's dirty and has got vibe! So the perfect treatment was to re-amp it to give it that vibe."

Figure 13-12. Keb' Mo'

224 RECORDING UNHINGED

Shelly Yakus on the Underwater Trojan Effect

"One day Roy Cicala from the Record Plant Studios in New York said to me, 'I want this singer to sound like he's underwater when he sings this one part,' so he said, 'Go to the drugstore and get a pack of rubbers—a pack of Trojans—and get a half gallon of milk, too.' So we poured the milk out of the glass bottle and we had one of these long, skinny microphones, and he put the rubber on it. And then we filled the milk bottle with water and we put the mic in the milk bottle. Right? So now the mic wasn't going to get ruined from the water, but it was in the milk bottle suspended in the water. Then we put a headset on the sides of the milk bottle and we sent the vocal in through the glass and listened. It sounded awful! So then we tried the same thing in an oil drum. We found a big metal drum they had there in the basement of the studio for cleaning fluid, and we taped headsets to the side of it. Then we dropped that same mic down inside with no water and that worked! That was for the group the McCoys that had that hit 'Hang On Sloopy.'"

---

upside down to get more or less of the snares rattling. You can really enhance the original snare by adding this into the mix. You can even compress and gate the room to create that kind of 'nonlinear' room sound for drums without having spent any money."

ROSS HOGARTH ON THE PASSIVE TRANSFORMER "IRON GIANT"

"Find yourself a standard passive transformer like a UTC or an RCA or a Malotki. (Passive transformers are the transformers that are in Fairchilds or Teletronix LA-2As.) Now, on the top of that transformer, there's going to be a pinout that is generally wired to go from 200 to 200 ohms. You'll see that it's labeled that way. Normal, nominal zero in, zero out, not causing any gain at all. But if you resolder the in and out of it, making the output now 50 ohms—which is a lot less resistance for that transformer—then the transformer itself starts working much harder and creates a tone. You will get amazing distortion from that transformer—really killer, beautiful distortion. You will really hear the metal working.

Figure 13-13. Ross Hogarth's "Iron Giant."

"How do you use it? Wire your transformer, lowered to a 50-ohm output, into the middle of an XLR cable. Then just take a track straight out of Pro Tools line out, into the transformer. Bring it up in a mix and you'll already hear what it does. You'll hear the sweet harmonic distortion of the metal. Now patch it into a compressor or some other processing! This is something I've been playing with, because I love the sound of passive transformers."

> ## Jack Joseph Puig on Mick Jagger's "Wedding"
>
> "When first working with Mick Jagger and the Rolling Stones, I explained to Mick that I like to have a day to listen and poke around the recording, creating the direction of the mix, then have the artist come back the following day at around four o'clock p.m. He said, 'Absolutely not. I won't do that.' Why not? 'Because if I come back the next day at 4 o'clock and I don't like the mix, I will have wasted all that money.' I was taken aback, but I looked at him and said, 'Well, if you want to stay, that's OK. You're the boss. It's your wedding.' This was a great lesson for me to not lose sight of whose 'wedding' it is—and Mick came back the next day at four o'clock anyway."

PAUL WOLFF'S "HEY-FIX" MIX CLARIFIER

"My nickname used to be 'Fix' back in the day, so I made this cool effect I used to call the 'Hey-Fix.' It's a single twenty-foot mic cable coiled up in a box with an XLR flush-mount connector input and an XLR flush-mount connector output. On the input, only the black wire is connected. On the output, only the white wire is connected. So when you put a signal into the box through a reverb send, the inductance and the capacitance of the black and the white wires will couple as they run next to each other through the twenty-foot cable. You'd come out of the box into a high-impedance input of a reverb return and blend the effect

Figure 13-14. Paul Wolff's "Hey-Fix" mix clarifier.

into your mix. This 'Hey-Fix' effect gives you 3 dB of high-end boost per octave! I used this effect during a Stéphane Grappelli concert to enhance the high end, and it made everything beautifully clear. Build two of them for clarifying a stereo mix!"

Ed Stasium on Talking Heads' Church Reverb

"When I did the first Talking Heads record, we recorded in a tiny little studio called Sundragon. We did the Ramones album *Leave Home* there, too. It had a little Roger Mayer console. The studio was tight, with no ambience whatsoever—twenty-five by twenty-foot little tiny room. No room for ambient mics. And when we were mixing at Media Sound, drummer Chris Frantz suggested that we make the drums bigger. This is on the album *Talking Heads 77*. Well, Media Sound was in a church, fabulous place. So we put some JBL 4310 speakers out in the room, pumped the kick and snare through them, miked the room with U 87s, then gated the return with Kepexes! You can really hear it on the song 'No Compassion.' The old church had sixty-foot-high ceilings! It was amazing! We did the same thing on later records at the Power Station studio in their large stairwell."

Figure 13-15. Talking Heads' 77 album.

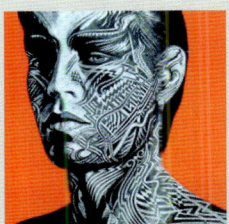

Figure 13-16. The Rolling Stones' *Tattoo You* album.

Bob Clearmountain on DIY Bathroom Reverb

"The reverb on the Stones' 'Start Me Up' was an extra bathroom (ladies', I think) in the basement of the Power Station (now Avatar), left over from when the building was a TV studio. I discovered it one day and noticed it had an interesting decay when I clapped my hands, so I cleaned it up, found an extra speaker—I think it was an old JBL 4311—and put up an X-Y pair of mics; can't remember what type, but they're probably still there. The studio techs wired it up to the studio's clever chamber network, so it could be accessed from any control room along with the rest of the plates and the magnificent seventy-five-foot stairway (R.I.P.). It was also the main 'verb on Springsteen's 'Hungry Heart.' Check out Danny Federici's Hammond solo in particular."

Figure 13-17. Stairwells make great reverbs.

Ed Stasium on the Ramones' Stairwell Reverb

"From 1977 well into the '90s, the Power Station in New York had a stairwell that was used for natural reverb. On the Ramones' record *Rocket to Russia*, it's the only reverb we used, because at the time the studio was so new, they didn't even have any digital reverbs. No plates either. I just used a little tape slap, and we used that stairwell on the east side of the building. I set up a crown DC-300 and two JBL 4310s, and I'd move the mics around and put everything on different levels. Eventually the Power Station had to stop using the stairwell because of fire regulations."

13-18. The Ramones with Ed Stasium.

BUILDING YOUR OWN COOPER TIME CUBE

The Cooper Time Cube was an early and completely unique attempt at an audio delay, designed by Duane H. Cooper and Bill Putnam. It was branded by Urei and released around 1971 as a studio effect. The sound of the Cooper Time Cube was very

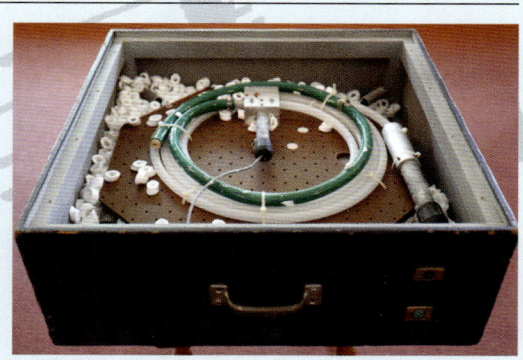

Figure 13-19. Inside the Urei Cooper Time Cube's mystery box.

Figure 13-20. DIY Cooper Time Cube.

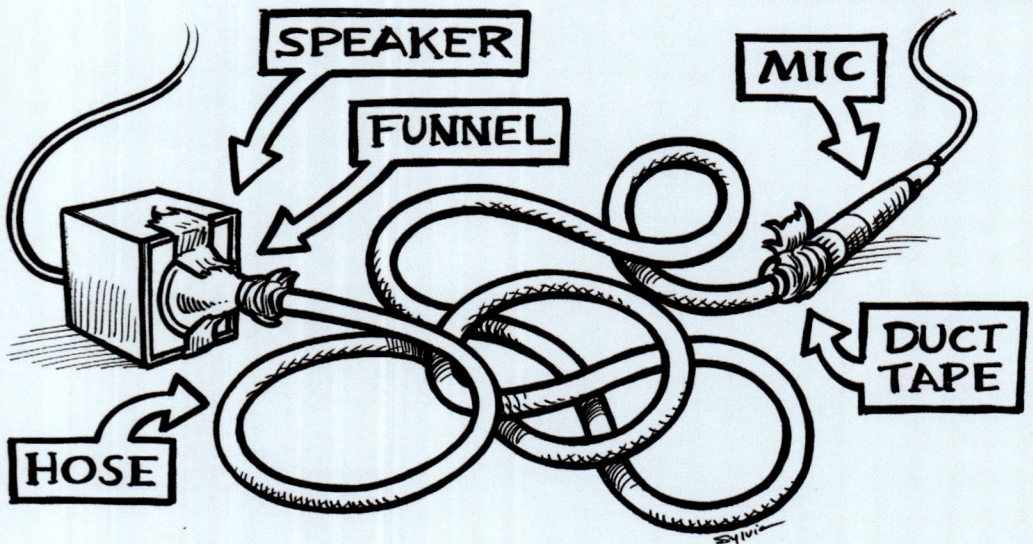

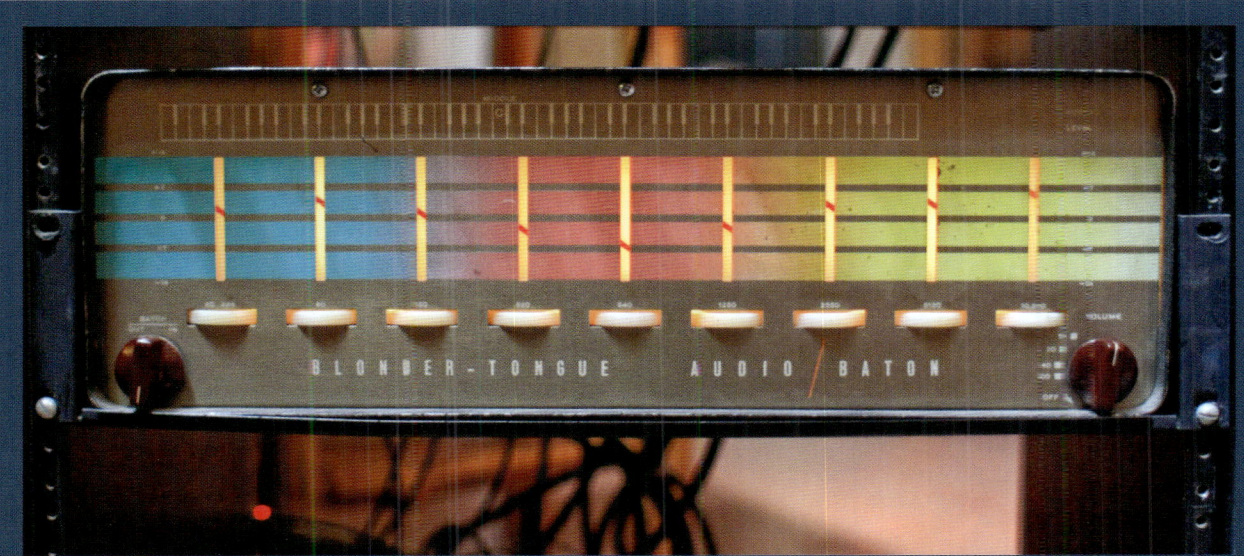

Figure 13-21. Blonder Tongue Audio Baton.

Ed Stasium on Awful Stereo EQs

"When I was working at Venture Sound I went to a garage sale and found not one, but two Blonder Tongue Audio Batons. Have you ever seen one of those things? Oh my god. It was like a cheap hi-fi thing. All the frequencies labeled on the front panel didn't sound like the actual frequencies at all. And it had little spinning things on it. And they lit up from behind with a rainbow of colors. As you turned a knob, a little dial would go up. They were amazing, but what shitty pieces of shit they were. Awful."

unusual, a short delay that kind of sounds hollow-ish. How does it work? Basically, sound is sent through a long hose. By the time the sound reaches the end of the hose, it has physically been delayed. Wow! Genius! An actual Cooper Time Cube unit consists of a rack-mounted "brain" unit and a separate large, mysterious box. I have one of these units and opened up the box to look inside.

The box contains an actual hose, similar to a garden hose. This hose has a small speaker on one end, with a small microphone picking up the sound on the other end. Hey, that doesn't look so hard to make! So I did. And I've included a diagram on how you can also build your own! It actually works great and can be customized to fit the tempo of the song you are using it on. Just use scissors to adjust the length of the garden hose!

TCHAD BLAKE ON TOM WAITS'S SPLATTERED MIX

"One of the most memorable moments in the studio for me was while mixing some songs at Prairie Sun Studios for the Tom Waits album *Bone Machine*. These words aren't exact, but it's how I remember it some twenty-odd years on.

"I had gotten to a place where the song 'Goin' Out West' seemed ready for playback to

Figure 13-22. Tchad Blake and his son Stan working in Full Mongrel Studios.

get Tom's comments. Tom sat while he listened at the desk, jerking his body about until the song ended. He seemed to think for a minute before he turned to me and said:

"'You know, Tchad, it all sounds great, really mean it. Love how the vocals and drums are sitting together and love the vibe overall. It's just the electric guitar—it sounds like it's not really part of the same trip. It sounds like it's splattered on the windshield when it ought to be in the backseat with little Billy and Brenda. The rest of the family's all in the car speeding down the road, with this big mess splattered there across the windshield. What I want to hear is everyone, back inside the car, screaming and laughing, all sitting together—one big happy family out for a drive. You get that and we're done with this song.'

"I've heard and read a fair amount of debate regarding appropriate communication in the studio over the years, with many people arguing for clear, scientific, or technical language while being frustrated with the bent and colorful. Maybe it's just me, but I can't recall a clearer set of instructions I ever had for anything that took all the stockholders into one account. Instruments, level, EQ, panning, effects, compression—all expressed in less than sixty very focused seconds, no questions asked. Another sixty seconds in adjustments and the mix was done.

"Not a year goes by I don't think about that moment at least a couple of times and learn from it."

Figure 13-23. Tom Waits.

Matt Wallace on Having More Faith in Our Own Abilities

"On that Faith No More album *The Real Thing*, I actually was going to quit producing the day I mastered that record. I was sitting there with mastering engineer John Golden, and the thing was so high-endy and had so much compression, and it just sounded dismal to me. And I listened on my car stereo, and it sounded like crap, and on my home stereo it sounded like crap. And I called my mom and asked her, 'How do I get into real estate? Because I have no idea what I am doing.' And I felt that way for a long time. But the moment it came on MTV, the song 'Epic' fucking blazed. And when it came on the radio, it sounded amazing! Technically, it's not that good-sounding a record. It's technically all wrong. But people love that energy—it's undeniable. I had a dbx 166, a cheap stereo compressor, and I ran Mike Patton's voice in one channel, then out into the other channel. Then I printed it. Then, during the mix, I took Patton's voice again into one channel of the same dbx 166, out of it into the other channel. Back into the console where we had bus compression, too! It was squashed to hell! It was awful, but man, it had energy, and it just jumped out of the TV speakers."

Ross Robinson on Knowing When You Are Done

"There will be a sense of completion inside of me—I'll wait for it. If I'm sitting and overanalyzing two bars of music, then I'm going to see a million problems that need to be fixed. But if I listen from the back, and listen all the way through, then I'll know if it is finished or not. I don't trust myself when I get into 'dog ear' mode. I will have to wait till the next day to check myself. When I get overly critical, it will never be perfect, but it will be *absolutely perfect* if I leave it the hell alone. Trust the feeling I have here in my core, rather than in my head."

WHEN IS A MIX FINISHED?

The end approaches. The marathon has been run, and now it's time to take the project over the finish line. The band listlessly wanders, eating all the Top Ramen and leaving ass creases in the furniture. They're busy plotting the wheelbarrow load of last-minute ideas they're about to dump all over you. Part politician, part marriage counselor—you must negotiate the tricky route toward the Promised Land: an actual day off!

But seriously, when is a mix truly done?

I have a bit of a reputation when it comes to mixing. I make my final decisions with confidence, and my intuition is strong.

When the song is done, it's done. I don't beat myself up.

Here are some words of wisdom for mixing and all else you do as an artist: "The project is not done when the ideas run out." In fact, the project is done long before the ideas run out, because as an artist, you have an oil well of ideas in your head that never stops flowing.

The trick is to understand when you've made your statement in a way that someone else can listen to and understand. That's when it's done—when it's a piece of art that stands on its own and conveys the snapshot of time that encompasses the studio, the people, the song. You must know that the river flows endlessly ,and your job is only to tote one bucket at a time.

Figure 13-24. Sylvia's world.

Figure 13-25. Sylvia Sassy!

ACKNOWLEDGMENTS

On those days when I find myself recording in a castle in Dresden or an island prison in Helsinki, setting up ridiculous musical contraptions for the day's sessions, I give a prayer of thanks. This joyful prayer is at the heart of this book, *Recording Unhinged*. It is an odd and wonderful thing we do, transforming ideas into audio legacy. It is fearless, and at the same time, it is a foolhardy quest. This book's purpose is to remind ourselves to not lose the sense of wonder that got us hooked in the first place. We give up stability, solid relationships, and financial security to leap bravely into the electrified swamp that is the recording business.

There are many of us on this path, and I'd like to take a moment to acknowledge those who generously donated their time to share their own unhinged stories, techniques, images, and knowledge to create this book, including Al Schmitt, Brian Malouf, Ed Stasium, Alan Meyerson, Garth Richardson, Linda Perry, Dave Pensado, Hans Zimmer, Elliot Scheiner, Paul Inder Kilmister, Geoff Emerick, Jeff Lorber, Bruce Swedien, Leland Sklar, Julian Colbeck, Michael Franti, Matt Wallace, Nick Launay, Ross Robinson, Susan Rogers, Bob Clearmountain, Brad Wood, Eric Valentine, George Drakoulias, Bob Ezrin, Ross Garfield, Peter Schickele, Jack Schumann, Joe Henry, Ross Hogarth, George Massenburg, Michael Beinhorn, Chris Shaw, Mark Rubel, Jack Joseph Puig, Jay Baumgardner, Paul Wolff, Marcella Araica, Ron Saint Germain, Damon Fox, Larry Crane, Tchad Blake, James Saez, Tim Palmer, Don Massy, Ida Moody, Adam Katz, Lori Castro, Ian Rickard, Justin Stanley, Shelly Yakus, Alex Kane, Jim Messina, Eric Anest, KaRIN, Mark Christian, and Emilio Solas.

I'd also like to express deepest gratitude to the folks at Hal Leonard Corporation for their support and patience through the writing and editing of this literary undertaking, in particular John Cerullo, Bill Gibson, and Lindsay Wagner.

And for their friendship, inspiration, and support, thank-yous go out to so many fellow travelers and troublemakers: Rick Rubin, Alan Parsons, Lisa Parsons, Ed Cherney, Rose Mann-Cherney, Chris Lord-Alge, Devin Powers, Sir George Martin, David Shiffman, Chris Goss, Josh Jerue, Matt Hyde, Herb Trawick, Chris Pelonis, Chuck Ainlay, Doug Fergus, Claudette and Michael Fitzgerald, Neil Portnow, Maureen Droney, Steve Smith, Jay Massy, Jamie Seyberth, Kimberly Freeman, Jason Sewell, Dave Bianco, Dina Ricker, Jeff Greenberg, Jeff Benson, Paul Diaz, Bailey Beechler, Dan Daley, Arno Jordan, Dan Workman, Tom Anker, Kevin Mills, Lauren Burke, Keith Olsen, Ben Folds, Danny Elfman, Pauley Perrette, Jeff Pevar, Michael Brauer, Jim Abdo, Michael Ross, Lisa Roy, Dani Macchi, Barry Rudolph, Eric Bonetti, Dusty Wakeman, Lenise Bent, Kai Huppunen, Matt Sorum, Danny Carey, Evan Davidson, EveAnna Manley, Bill Hahey, Leslie Ann Jones, Michael Romanowski, Bill Kaylor, Dave Smith, Michael Zähl, Jim Keltner, Frank Black, Mike Clink, Craig Chaquico, Danny McGough, Cliff Chang, Lisa Haley, Michael Wagener, Frank McDonough, Rich Hansen, Mike Broman, Darrick Bob Jones, Dave Aron, Andrew Scheps, James McKinney,

Kale' Holmes, Mary Mazurek, Josh Freese, Abbey Loso, Bob Miner, Matt Hill, Jim Spiri, John Cuniberti, Jimmy Douglass, Zachary Vex, Mark Arinsberg, Andy Wallace, Tina Rose Day, Maor Appelbaum, Joe Haze, Mike Lawson, Glenn Lorbecki, Gary Boss, Dave Hudson, Clare Pproduct, Brian Kehew, Dave Trumfio, Jim Scott, Jon Phelps, Mike Fisher, Matthew Kriemelman, Dave Reitzas, Luana Caraffa, Traditional Medicinals, Brent Turner, Thomas Dolby, Mark Pauline, John Taylor, Phil Spector, John Jones/Thumper Drums, Dave Way, Howard Massey, Ziggy Modeliste, Greg Frederick, Hugh Padgham, Paul Rivera Jr., Dave Storie, Mr. Bonzai, Rupert Neve, Joshua Thomas, Candace Stewart, Carmen Rizzo, Dave Gross, Tom Kenny, Trina Shoemaker, Holly and Jim Connell, Ernie Ball, Tony Visconti, Wesley Bulla, Ron Nevison, Martin Weischermann, Rafa Sardina, Joseph Crowell, David Gelfand, Charlie Steves, Mitch Easter, Phil Moore, Bob Ludwig, Roxanne Ricks, Scott Van-Fossen, Paul D'Amour, Vicky Giordano, Louis Smith, Mike Matthews, Serj Tankian, Robert Margouleff, Murial Boyer, Tim Stearns, Joe Chiccarelli, Charlie Paakkari, Robert Wheeler, Jason Gallagher, Øystein Greni, Ross Heinemann, Sarah Burns, Brendan Duffey, Dave Watts, Tom Stamper, Shanine Yngvason, Rory Kaplan, Jeff Stanley, Peter Junge, Sherri Tantleff, Jimmy Boyle, Morgan O'Shaughnessey, Richard and Lindsay Pursel, Terry Manning, Pat Metheny, Marco B, Paula Salvatore, Chris Dauray, David Bock, Nick Kirby, Shawn Montgomery, Sarah Jones, Dave Hampton, Wes Dooley, Mike Mathis, Rich Rees, Rich Veltrop, Piano Showcase Medford, and Norwood Fisher.

For anyone that takes the time to enjoy this book, an extra special THANK YOU.

Live long, and make every recording session an adventure!

INDEX

accidents, 5–7
Adams, Bryan, 36
ADSR (attack, decay, sustain, release), 170–71
After Bathing at Baxter's, 195
Aguilera, Christina, 65
AIR Studios, 34–35
Aitken, Doug, 39–41, 175
Aja, 6
Albini, Steve, 31, 83
Alpenhorn, 162
Ampex 300, 47
Ampex tape machines, 48–49
amplifiers, 14
    Acoustic "Tuned Tube," 125
    Ampeg, 10, 90, 93–95
    "Barbie," 129
    Collins 26U limiting, 26
    design for no, 141
    Marshall, 123, 149
    microphones and, 132
    "micro twin" Fender, 128
    RadioShack, 129
    split-amp technique, 132–33
    Teisco tabletop, 125
    3rd Power, 125–26
    tiny and toy, 128–29
    Tone Tubby, 125–26
analog recording, 16–17, 48–49
anger, vocals and, 79
Anker, Tom, 59
Aphex 602 processor, 209
API, 20–21
Araica, Marcella, 200
Armstrong, Dan, 125
Artis the Spoonman, 178
attack, decay, sustain, release. *See* ADSR
Aubert, Dani, 83
auto-tune, 67

Babes in Toyland, 31, 201
Bad Brains, 51
bagpipes, 161, 163
Baker, Roy Thomas, 195
Band of Horses, 42–43
the Bangles, 218
Bangsburg, Rayne, 57
bass
    DI boxes for, 90
    Dingwall, 95
    distortion and, 92–93
    Electro-Voice RE-20 for, 91
    feeling, 97
    Fender Jazz, 90

    gloves, 89
    groove of, 89
    on "Lady Madonna," 94
    mixing, 222
    re-amping, 222–23
    recording-chain for Tool, 91
    rigs, 90–96
    SansAmp units for, 93
    sexy, 89–90
    six string, 96
    strings, 95
    upright, 97, 154
    Warm Audio EQPWA for, 91
Baumgardner, Jay, 63
Beach Boys, 178
beaters, 107–8, 111–12
the Beatles, 11, 83, 94, 188, 190, 193, 195, 201, 211, 213, 217
Beatnigs, 181–82
Bechirian, Roger, 135
Bee Gees, 68
Beinhorn, Michael, 26, 30, 86, 178
Bell, Vincent, 124
Bellybutton, 110
Benson, George, 66
"Billie Jean," 85
Billion Dollar Babies, 36
Bjelland, Kat, 201
Black, Frank, 46
Black Crowes, 193, 201, 203
Black Hawk Down, 158–59
Blackmore, Ritchie, 94
Black Sea, 212
Blake, Tchad, 229–30
bleed, 10, 12–13, 105, 111–112, 115, 192
Bock, David, 75
Boddicker, Michael, 167
Borich, Lucius, 103
Bowie, David, 74, 81
Boyle, Jimmy, 24
Breezin', 66
Brides, 184
bridge, 207
Bridge of Sighs, 14
Brill, Dusty, 18
Brook, Michael, 163
Browne, Jackson, 41, 53
Brownstein, Carrie, 78
Bruford, Bill, 175
building equipment, 140–41
Burl, 49
Burnett, T-Bone, 113
Bush, Kate, 112

Cage, John, 145, 190, 209
Caraffa, Luana, 74
Carey, Danny, 52, 103, 221
Carnegie, Scott, 124
Carter, Betty, 47
Cash, Johnny, 24–25, 125, 147, 199, 204
Castle Röhrsdorf, 32
Castro, Lori, 162
Cave, Nick, 182
Charles, Ray, 47
Chavez, Ingrid, 217, 221
Christian, Mark, 135
Cicala, Roy, 113, 225
Clearmountain, Bob, 36, 42–43, 219, 227
Clearwell Castle, 32
click track, 203–4
Clooney, George, 104
Coal Chamber, 47
Coffeen, Tom, 15
Colbeck, Julian, 4, 155, 168–69, 174–75
Coleman, Lisa, 21
compression, 23–24, 27, 31, 51, 71–72, 231
compressors, 24–31, 51, 80
Computer World, 198
consoles, 16–23
Content, Filipe, 39
control room, 138–40
Cooke, Sam, 57
Cooper, Duane H., 228–29
Coral sitar, 124
Corsaro, Jason, 178–79
"Cosmic Beam," 183
Costa, Nikka, 77
Countryman DT-85 DI boxes, 90
Coyne, Wayne, 38
Crane, Larry, 13, 45, 78, 127, 130, 171, 219
"Crawl Away," 79
Cuniberti, John, 222–23
the Cure, 95
cymbals, 105–6, 111–112, 118

D'Amour, Paul, 90
Davis, Jonathan, 61, 163
Davis, Matt, 139
"Dear Jane," 158
DeBaun, Jon, 1
Decca Tree, 155–57
Deftones, 199
delay, 204
depth of field, 155

"Desire," 169
Diamonds and Pearls, 48–49
DI boxes, 90, 92, 222–23
Dies, Josh, 172
distortion, 23, 92–93, 124
dogs, 54–55
the Doors, 221
double compression, 23–24
Dragonfly, 105
Drakoulias, George, 15, 149, 201, 218
The Dreaming, 112
drums, 11. *See also* cymbals; percussion and noise
  alternate ideas for, 120–21
  beaters, 107–8, 111–12
  click track, 203–4
  cocktail, 119
  damping, 112–13
  drummers, 100–101, 112, 203–4
  dynamics, 112
  EQ, 101
  as foundational, 99
  groove, 100
  heads, 102–4, 108, 112–13
  kick, 105, 107–13
  Ludwig, 113–14
  machines, 105, 110, 138, 203
  material of, 114
  microphones, 101, 108–9, 111–12, 115–16, 118
  MIDI, 175
  monitors, 108
  PA systems, 116
  percussion and noise, 15, 51–53, 177–85
  phase and, 108
  position of, 101
  prepared, 120–21
  reverb for, 112
  rugs, 101
  in small rooms, 117
  snare, 83, 99, 101, 108, 111, 113–15, 119–20, 223–25
  sound of, 99, 101–2, 112, 116, 118
  Spaun, 119
  sticks, 107
  Tama, 114
  tambourine snare, 120
  "Terminator" snare, 114
  toms, 101, 103–5, 111, 221–22
  Trixon, 118–19
  tuning, 102–4
  vocals into snare, 83
  weird, 118–19
duduk, 163

the Eagles, 41
Early Electronic and Tape Music, 190
Earth, Wind, & Fire, 159
Ebow, 147
Edges of Twilight, 79
Electronium, 189
Ellis, Warren, 182
Emerick, Geoff, 11, 14, 80, 136, 155, 201, 211, 213
End of the Century, 129
Eno, Brian, 198
environment, 78–79. *See also* spaces
EQ, 16, 47–48, 78, 101, 133–34, 137, 216, 229. *See also* consoles
equalizers, 22. *See also* consoles
equipment, 3, 13, 15–16, 24–26, 140–41
Erb, Werner, 162
Euphonic Audio, 92
EV 666, 13, 130
"Even Here We Are," 12
evolution of recording, 16
Ezrin, Bob, 36–37, 42, 61, 187, 204, 213

Faith No More, 206, 231
Farmhouse, 37
feedback, 15, 138–40
Ferrone, Steve, 113
Fetchin Bones, 21
Finn, Tim, 69
Fisher, Eddie, 195
.5: The Gray Chapter, 112
Flavor Flav, 81
Flea, 97
Flowers of Romance, 161
Folds, Ben, 144
Fornax Chemica, 216
4 Non Blondes, 87
"Four Enclosed Walls," 161
Fowler, Bruce, 159
Fox, Damon, 49, 169, 173
Fragile, 91
Frampton, Peter, 138
Frampton Comes Alive!, 138
Frank Furter and the Hotdogs, 45
Franti, Michael, 181
Frederick, Greg, 95
"Free Fallin'," 127
The Friends of Rachel Worth, 219
From First to Last, 34
Furtado, Nelly, 200

Gabriel, Peter, 22, 37–38, 61
Gadd, Steve, 195
Galesi estate, 36
Garfield, Ross, 102, 104, 112–14, 117
Gaugh, Bud, 120–21
gear. *See* equipment

Geer, Didrik De, 157
Geggy Tah, 53, 82, 100, 144
Gibson, David, 220
Gilmour, David, 42
Gilt, 14
Girlfriend Experience, 99, 202
Girls, Girls, Girls, 116
Gladiator, 163
Go-Betweens, 219
"Goin' Out West," 229–30
Golden, John, 231
Good Charlotte, 18
"Good Times," 191
Gotye, 166
Greetings from California, 158
Gregory, Cecil, 90, 127
Greni, Øystein, 28
groove, 100
guitars, 123. *See also* amplifiers
  attitude and, 123
  backwards, 211
  baritone, 124
  cliff-toss of, 15
  EQ, 133–34, 137
  feedback, 138–40
  feet of guitarists, 136
  hairdryers for fuzz effect on, 135
  ham-fisted guitarists, 127
  through Leslie speakers, 134
  microphones and, 130
  Nashville style of stringing, 127
  pedals, 135–38
  phase and, 130–31
  relearning to play, 128
  strings, 127
  surface transducer, 139–41
  sustainer, 140
  tape-machine distortion for, 124
  tuning, 127
  Variac units, 126

"The Hammer," 47–48
Hammond organs, 148–50
"Hang On Sloopy," 225
Happy Days, 213
harmony, 68
harp, 162
harpsichords, 148
Harrison, Andy, 128
Harvey, PJ, 177
Have One on Me, 162
"Hazy Shade of Winter," 218
Headley Grange, 35
headphones, 72–74, 184
"Helicopter String Quartet," 191–92
Henley, Don, 205
Henry, Joe, 183
Heroine, 34
high-pass filtering, 110
Hill, Matt, 136
Hogarth, Ross, 53, 65, 84, 109, 126, 180, 184, 200, 205, 225

Hohner Clavinets, 148
Hold Out, 53
hooks, 206
Horn, Paul, 32–33, 161
horns, 153, 155, 159
Höskulds, Husky, 183
"Hot Thing," 138
Hot Tuna, 210
Houston, Whitney, 6
Howard, Ron, 34
Howlin' Maggie, 128
H. R., 51
Hurt, 99
Hutcherson, Bobby, 115

I Against I, 51
Igudesman, Aleksey, 158
Indigo Ranch, 133
industrial art movement, 181
Interstellar, 1, 34, 151
In Through the Out Door, 32
Into the Fire, 36
intros, 201
iPhone, 49–50, 138
"Iron Giant" passive transformer, 225
"I Will Always Love You," 6

Jackson, Michael, 85
Jackson 5, 70
Jagger, Mick, 46, 62, 226
Jazz Suite, 32–33
Jean, Norma, 34
Jefferson Airplane, 195, 210
Jefferson Baroque Orchestra, 156
Jellyfish, 110
Johns, Andy, 35, 116
Johns, Glyn, 193
Johnson, Chris, 51
Jones, Adam, 133
Jones, Quincy, 192
Jones, Spike, 190
Jordon, Tommy, 53
Joydrop, 82

Kaiser's Orchestra, 1
Kamankesh, Parisa, 77
Karno, Fred, 42
Katayama, Suzie, 158
Keb' Mo', 224
Keenan, Maynard James, 79–80
Keltner, Jim, 118, 180
kick pedal, 111
Killing Joke, 38
Kilmister, Paul, 113
Kjølsrud, Agnete, 62
Klepht, 89
Knifonium, 174
Knight, Gladys, 7
Korn, 61, 163
Kraftwerk, 198

Krauss, Vik, 97
Kurstin, Greg, 144

"Lady Madonna," 94
Landes, Scotto, 128
Lanois, Daniel, 33
Launay, Nick, 5, 112, 161, 166, 182, 212
leakage, 156
Leatherheads, 104
Led Zeppelin, 32, 35, 116
Lee, Tommy, 116
Left, Jeff, 99
Lennon, John, 80, 83, 156, 193, 211
Leslie speaker, 83, 134, 149
"Like a Feather," 77
limiters, 27–29
Lindley, David, 53
Linn, Roger, 173
Little, Jonathan, 223
live feeling, 12
live-to-two-track, 45–46
Lorber, Jeffrey, 169
Lord, Jon, 149
Lord-Alge, Chris, 218
Love, Courtney, 13
low-frequency, 110
Lydon, Johnny ("Johnny Rotten"), 161
Lynch, Stan, 184, 205
lyrics, 47, 57, 59, 202, 206

MacDonald, Phil, 195
Machines of Loving Grace, 14
MacLeod, Brian, 105
The Madden Brothers, 158
Malouf, Brian, 6, 192
Mankind Is Obsolete, 128
Manning, Roger, 143
Man of Steel, 117
marijuana, 87
Marius, 12
Markley, Dean, 141
Marley, Ziggy, 105
Maroon 5, 29
Marshall, Jim, 94
Martin, George, 34–35, 188
Martin, Jeff, 79
Massenburg, George, 159
Massy, Don, 86
Mayer, Roger, 28
the McCoys, 225
McKee, Maria, 15
mechanical revolution, 165
Mellencamp, John, 200
Menlove Ave, 156
Messina, Jim, 47, 221
Metheny, Pat, 166–67
Meyerson, Alan, 158, 183
Mickey & Sylvia, 191
mic pres, 22–23, 78, 141
microphones, 71. *See also* phase

AKG C1000, 79–80
amplifiers and, 132
Audio-Technica, 111, 162
Beyerdynamic M 160, 35
bleed, 10–12–13, 105, 111–112, 115, 192
blending, 131–32
Blue Hummingbird, 145
Decca Tree, 155–57
drums 101, 108–9, 111–12, 115–16, 118
Electro-Voice RE-20, 91, 109
EV 666, 13, 130
guitars and, 130
handheld versus suspended, 80
hanging, 115–16
Mojave MA-100, 155–56, 184
muddy, 130
MXL 990, 75
Neumann M 49, 116
Neumann U 57, 75, 80, 84
O.A.R. 75
orchestras and, 155–58
pianos and, 144–45
positioning, 111, 115–16, 134
power hazards, 82
Pressure Zone or PZM, 77, 118, 135, 145
RCA BK-1A "Ice Cream Cone," 77
ribbon, 130, 157–59
Sennheiser 441, 81
Shure 555 130
Shure Beta 181, 178
Shure SM57, 75, 81, 118, 158
Shure SM58, 34
sub, 109–10
summing, 131
sweet spot for, 134–35
technique, 31
Telefunken U 47, 75, 82
telephone, 35–86
tube, 157
vintage, 76
vocals, 75–77, 80, 84
MIDI recording, 175
"Midnight Train to Georgia," 7
Mighty Lemon Drops, 134
M.I.K.E. (Musically Integrated Kiosk Environment), 41
mixing, 201
  adventure, 217, 221
  bass, 222
  clarifiers, 226–27
  client relationships and, 226
  by committee, 217
  completion of, 231
  digital, 219
  long-distance, 219
  modern, 220
  panning, 221–22
  perfection and, 215–17, 231
  re-amping with "Double DI" technique, 222–23

repair, 216
Studio Technology AN-2, 222
thinning out, 218
visual concepts for, 220
vocals, 222
Waits and, 229–30
"Modern Love," 61
monitors, 12, 108, 184
Monkees, 135
Moody, Ida, 64
Moon, Keith, 193
Moore, Phil, 25
Moore, Sonny. *See* Skrillex
Moore, Thurston, 40–41
Mothersbaugh, Mark, 189
Mötley Crüe, 116
motorcycle solo, 185
Muscle Shoals, 149
music, 67
musical flavors, 207–8
Musically Integrated Kiosk Environment. *See* M.I.K.E.
MXL 990, 75
My Own Private Alaska, 209

Nashville-style, 127
negative space, 209
Nelson, Willie, 33
Neumann U 67, 75, 80, 84
Neumann U 87, 51
Neve consoles, 16–19, 23
Nevercore, 95
Nevermind, 114
New Beat Fund, 97
Newsom, Joanna, 162
Nicks, Stevie, 63
Nirvana, 114
Nolan, Christopher, 1, 50, 151
"No Woman, No Cry," 150
nyckelharpa, 160

"Oblique Strategies," 198
"Oceans," 180
Offspring, 206
orchestras, 153–58
orchestrion, 165–67
organs, 148–51
O'Shaughnessy, Morgan, 160
outros, 201
outtakes, 5
overproduction, 213

Page, Jimmy, 35
Palmer, Tim, 74, 81, 134, 180
panning, 221–22
Panunzio, Thom, 218
Papa Roach, 63
"Parking Lot Experiment," 38
passive transformers, 225
PA systems, 12, 84, 116

Patchy Sanders, 83, 204
Patton, Mike, 199, 231
Paul, Les, 192
Paul, Stephen, 82
Pauline, Mark, 181
P.D.Q. Bach, 107
"Peace for the Wicked," 62
Pearl Jam, 180
Peart, Neil, 120
Penn, Dan, 149
Pensado, Dave, 215
percussion and noise, 15, 51–53, 177–85
perfection, 66–67, 215–17, 231
Perrette, Pauley, 79, 158
Perry, Lee ("Scratch"), 194
Perry, Linda, 13, 23, 58, 63, 65, 73, 78, 87, 196–97, 215
Peter Gabriel 1, 61
Peterson, 167
Pet Sounds, 178
Petty, Tom, 113, 127, 199
phase, 108, 130–31
Phish, 37
pianos
    dropping, 52
    with Ebow, 147
    electric, 147–48
    extended techniques for, 146
    Fender Rhodes, 147–48
    grand, 144
    harpsichords, 148
    Hohner Clavinets, 148
    Jaymar toy, 120–21
    lid reflections of, 145
    microphones and, 144–45
    orchestras and, 155
    organs, 148–51
    prepared, 146
    shooting, 52–53
    single-note impact, 147
    size of, 144
    smashing, 53
    string, 146
    tack, 146–47
    upright, 143, 145–46
    Vari-speed on, 144
    Wurlitzer electric, 147–48
Pink, Ariel, 41
pipe organs, 150–51
Pirner, David, 10
the Pixies, 46
Plank, Konrad ("Conny"), 198
planning, 5
Pluto's Cave, 40
Pool, Fred, 136
pornography, 47
"Power Fantastic," 21
power hazards, 82
power tools, 182
Pproduct, Clare, 23
preciousness, 4
preproduction, 200–201

Presidents of the United States of America, 95
pressure, 202, 209
"Pretty Fly (For a White Guy)," 206
Prince, 21, 48–49, 70, 138, 196, 221
producers
    breakthrough ideas for, 211
    independent, 193
    mad scientist, 199
    musician, 195–96
    negative space and, 209
    overproduction by, 213
    pioneering, 189
    rebel, 197–98
    relationships of, 188
    role of, 187–89
    setting limits, 209
Pro Tools, 131, 204, 218
psychology, 205
Public Enemy, 81
Public Image Ltd, 161
Puig, Jack Joseph, 2, 110, 193, 203, 226
Putnam, Bill, 228–29

RadioStar Studios, 33–34, 70, 162, 217
Rage Against the Machine, 11
"Rain," 211
Ramone, Phil, 194
Ramones, 123, 129, 228
"Rapper's Delight," 191
Rattle and Hum, 169
The Real Thing, 206, 231
Real World Studios, 37–38
re-amping, 222–25
Record Plant, 9, 28
Redeemer, 34
"Release the Bats," 182
"Respectable Street," 212
reverb, 78, 112, 223–25, 227–28. *See also* spaces
Revolver, 11, 211
ribbon microphones, 130, 157–59
Richards, Keith, 125
Richardson, Garth, 11
Rickard, Ian, 40, 45
Robinson, Ross, 4, 7, 33–34, 43, 61, 79, 82, 86, 96, 133, 163, 199, 204, 209, 231
Rocket to Russia, 228
Roger Mayer RM58 limiter, 28
Rogers, Susan, 12, 21, 49, 53, 69, 82, 100, 144
Rolling Stones, 93, 226–27
Roma music, 208
Roots, 43
Rose, Morgan, 100
rotary speakers, 134
Roth, Daniel, 150
Roth, Marius, 137
Rubel, Mark, 107, 121
Rubin, Rick, 13, 71, 84, 147, 199, 222

Sacred Cow, 53
"Sacred Love," 51
SansAmp units, 93, 215
Satriani, Joe, 19–20
Saturday Night Fever, 68
Sayer, Roger, 151
Scar, 183
Scarecrow, 200
Scheiner, Elliot, 6, 9–10, 69, 194
Schickele, Peter, 107
Schmitt, Al, 22, 31–32, 47, 66–67, 161, 188, 192, 195
Schumann, Jack, 160
Scott, James, 125
Scott, Jim, 113
Scott, Raymond, 189–90
Scott, Ridley, 163
self-confidence, 65
Sepultura, 43
shakers, 180
Shake Your Money Maker, 201
Shannon, Del, 49
Shaw, Chris, 81
Sherlock Holmes, 208
Sherman Filter Bank, 109
"She Withers," 53
Showbread, 139
Siddhartha, 86
Sign 'O' the Times, 138
Sihasin, 162
Sinatra, Frank, 67
singers, 58. *See also* vocals
The Sixth Sense, 217
Sklar, Leland, 92–93
Skrillex (Sonny Moore), 34
Skunk Anansie, 204
Slash, 136
Sleater-Kinney, 78
Slipknot, 112, 114
small rooms, 117
Smashing Pumpkins, 13
smashing things, 51–53
Smith, Robert, 96
SOAD, 71
"Sober," 134
Solid State Logic (SSL) consoles, 22
Solla, Emilio, 208
Sonatas and Interludes, 146
Songs About Jane, 29
song structure and content, 200–201
Soul Asylum, 10
Soundgarden, 178–79
spaces, 31
    barges, 42
    canyons, 43
    castles, 32
    caves, 40
    churches, 35, 227
    Egyptian pyramids, 38
    farmhouses, 37
    houses and mansions, 35–36
    mountains, 42–43
    parking lots, 38–39
    silos, 41
    small rooms, 117
    sweet spot in, 134–35
    temples, 32–33
    theaters, 34–35
    trains, 39–41
spark shooter, 181
Spector, Phil, 129, 156, 190
Spiderbait, 128
split-amp technique, 132–33
"Spoonman," 178–79
SSL. *See* Solid State Logic consoles
Stanley, Justin, 39–41, 77, 96, 141, 175
Starr, Ringo, 11
"Start Me Up," 227
Stasium, Ed, 3, 7, 10, 21, 46, 62, 68, 75, 79, 123, 130, 135, 162, 195, 227–28
"Station to Station" studio, 39–41, 175
Steely Dan, 6
Steigmeyer, Wade, 216
St. Germain, Ron, 38, 51, 82
Stockhausen, Karlheinz, 146, 191–92
strings, 153, 161, 183. *See also* upright bass
Stroh Violin, 161
studio posses, 63
Studio Technology AN-2, 222
Sublime, 90, 120–21, 174
sub microphones, 109–10
"Such a Beautiful Night," 144
Sugarhill Gang, 191
Sundet, Per Kristian, 9
Superunknown, 178–79
surface transducer, 139–41
Surfing with the Alien, 20
Swan, Sierra, 63
Swedien, Bruce, 55, 75, 85
"Sympathy," 78
synthesizers, 167–75
System of a Down, 71, 84, 147

"Take It Easy," 41
talkbox, 138
Talking Heads, 227
Tankian, Serj, 71, 84
Tanner, Geoff, 18
Tattoo You, 227
Taylor, Phil, 42
Teatro, 33
Telefunken U 47, 75, 82
telephone microphones, 85–86
Temple Church, 34, 151
tempo, 203–4
Ten, 180
Tench, Benmont, 149
Their Satanic Majesties, 93
Thin Red Line, 183
"This Masquerade," 66
Thomas, Chris, 198
Thomas, Pete, 177
Three Snakes and One Charm, 203
Thriller, 85
Timbaland, 200
"Tomorrow Never Knows," 11, 83
toms, 101, 103–04, 111, 221–22
Tonight Alright, 128
Tool, 52, 79, 80, 84, 90, 91, 103, 133, 134, 221
Townshend, Pete, 94
transgender vocals, 85
Trident consoles, 19–21
Trower, Robin, 14, 136
tube microphones, 157
TubeWorks DI boxes, 92
Tucker, Corin, 78
tuning, 102–4, 127

U2, 169
Unchained, 24–25, 125, 147, 204
Undertow, 52, 79–80, 84, 103, 133
Underwater Trojan Effect, 225
upright bass, 97, 154
Urei Cooper Time Cube, 228–29

Valentine, Eric, 18, 25, 30, 48–49, 124, 132, 136, 158, 209, 220
Vanderpool-Robinson, Sylvia, 190–91
VanFosser, Scott, 2
Van Halen, Eddie, 126
Variac units, 126
Vari-speed, 49, 144
Veltrop, Rich, 217
Veruca Salt, 73
Vex, Zachary, 135
vintage tape sound, 49
violins, 161
vocals, 57
    ad-libs, 65
    aggressive, 32
    alcohol and, 85–87
    anger and, 79
    audience for, 70
    auto-tune, 67
    backwards, 212
    bad advice for, 87
    being on display and, 70
    blown-out, 25
    booth, 71
    compression for, 71–72
    connecting to moment in, 58
    from couch, 84
    devolution of, 81
    distraction techniques for, 59–60
    emotion and, 59
    environment and, 78–79
    EQ, 78
    through fan, 83
    friends included in, 63
    harmony, 68
    headphones and, 72–74

heroic, 58
ice cream truck, 59
as if onstage, 62
laziness of singers, 69
through Leslie speaker, 83
levels for, 73
lung power, 71
microphones, 75–77, 80, 84
mixing, 222
mumbling, 64
naked singing of, 62
PA systems and, 84
perfection and performance in, 66
pitch, 72–73, 86
posture, 64
reverb, 78
role of singer for, 62
sickness of, 69–70
singing into wall, 77
slacker technique for, 87
smiling and, 64
into snare drum, 83
sophisticated technique for, 87
sphincter and, 64
in straitjacket, 63
techniques, 64–65
transgender, 85
tube, 85
underwater, 82
upside down, 84, 232
vocal aids and remedies, 70
warm ups, 59, 79

Waits, Tom, 229–30
Wallace, Matt, 3, 12, 16, 30, 58, 67, 75, 87, 93, 95, 97, 127, 135, 140–41, 147, 206, 218, 223, 231
Wall and Bridges, 193
Warhol, Andy, 146
Waronker, Joey, 107
Watts, Dave, 101
Weinberg, Jay, 112
Westerberg, Paul, 12
Western Electric 1217 limiter, 27–29
"When the Levee Breaks," 35, 116
Wildflowers, 113
Wilson, Brian, 31, 178
Wilson, Eric, 90, 174
Wolff, Paul, 21, 109, 116, 118, 226
Wood, Brad, 48, 73, 117, 167, 207, 218
Wood, Jim, 128
Wyman, Bill, 94

Yakus, Shelly, 28, 36, 61, 63, 71, 94, 99, 101, 115–16, 156, 159, 193, 217, 219, 225
"Yellow Submarine," 213
Yes, 91

Zimmer, Hans, 1, 3, 34, 49, 117, 151, 158–59, 163, 167, 174–75, 183, 208

ILLUSTRATION CREDITS

1-1, 2-1, 2-11, 2-12, 3-1, 3-11, 4-1, 4-5, 4-12, 4-23, 4-42, 4-52, 5-1, 5-6, 6-1, 6-4, 6-22, 6-25, 6-33, 7-1, 7-24, 7-37, 7-40, 7-41, 8-1, 9-1, 10-1, 11-1, 11-14, 12-1, 13-1, 13-10, 13-11, 13-13, 13-14, 13-20, 13-24: Illustrations by Sylvia Massy

1-2, 1-3, 1-4, 1-5, 1-6, 1-8, 1-9, 2-6, 2-9, 2-10, 2-13, 2-15, 2-16, 2-22, 2-23, 2-25, 2-26, 2-27, 2-28, 2-29, 2-30, 2-32, 2-33, 2-35, 2-37, 2-38, 2-44, 2-51, 3-2, 3-4, 3-6, 3-7, 3-8, 3-10, 3-12, 3-14, 3-15, 4-2, 4-3, 4-4, 4-8, 4-9, 4-13, 4-20, 4-22, 4-27, 4-28, 4-29, 4-32, 4-33, 4-34, 4-35, 4-38, 4-40, 4-41, 4-44, 4-45, 4-46, 4-48, 4-49, 4-50, 5-2, 5-3, 5-4, 5-5, 5-10, 5-11, 5-14, 5-15, 5-16, 5-18, 6-2, 6-3, 6-5, 6-6, 6-7, 6-8, 6-9, 6-11, 6-13, 6-14, 6-17, 6-20, 6-21, 6-23, 6-26, 6-27, 6-28, 6-29, 6-30, 6-31, 6-34, 6-36, 6-37, 6-39, 6-40, 6-41, 7-2, 7-3, 7-7, 7-8, 7-9, 7-10, 7-11, 7-13, 7-15, 7-16, 7-18, 7-19, 7-20, 7-22, 7-23, 7-25, 7-26, 7-27, 7-28, 7-29, 7-30, 7-32, 7-34, 7-38, 7-39, 8-2, 8-3, 8-5, 8-6, 8-7, 8-11, 8-12, 8-13, 8-14, 9-3, 9-4, 9-5, 9-9, 9-11, 9-13, 9-14, 9-19, 10-5, 10-6, 10-7, 10-8, 10-10, 10-11, 10-12, 10-13, 10-14, 10-15, 10-16, 10-17, 10-18, 11-6, 11-12, 11-13, 12-8, 12-10, 12-13, 12-14, 12-17, 12-20, 12-22, 12-23, 12-26, 12-28, 13-3, 13-4, 13-7, 13-8, 13-9, 13-17, 13-19: Sylvia Massy's collection

1-7, 8-15: Photo courtesy Fox Faehrman

1-10, 2-8, 2-36, 2-43, 2-45, 2-49, 2-50, 2-52, 2-58, 3-9, 3-13, 4-7, 4-16, 4-37, 4-47, 4-51, 5-7, 5-13, 5-17, 6-12, 6-24, 6-32, 7-35, 8-8, 9-6, 9-10, 9-12, 11-3, 11-5, 11-7, 11-15, 12-5, 12-11, 12-12, 12-16, 12-18, 12-19, 12-25, 12-29, 12-30, 13-2, 13-15, 13-16: Chris Johnson's collection

1-11: PH2 Mark Kettenhofen/Wikimedia Commons

1-12, 2-3, 3-3, 4-10, 7-17, 7-21, 9-17, 13-18: Photo courtesy Ed Stasium

2-2: Photo courtesy Stefan Witts

2-4: Claudio Divizia/Shutterstock.com

2-5: Photo by Jeff Sheehan

2-7: Featureflash/Shutterstock.com

2-14: Brillzile/Wikimedia Commons

2-17, 2-18: Photo courtesy Paul Wolff

2-19: Nick Genin/Wikimedia Commons

2-20: Joi Ito/Wikimedia Commons

2-21, 7-12, 7-14: Photo courtesy Joshua Weinfeld

2-24: Photo by Brian Woodcock

2-31: Photo by Laura Grier, lauragrier.com

2-34: Photo courtesy Roger Mayer

2-39: Brother Records/Wikimedia Commons

2-40: AH Films/Wikimedia Commons

2-41: Photo courtesy Arno Jordan

2-42: Turtix/Shutterstock.com

2-46: John Salmon/Wikimedia Commons

2-47: Steve Cadman/Flickr Commons

2-48: Andrew Smith/Wikimedia Commons

2-53, 10-20: Photo courtesy Doug Aitken, Station to Station

2-54: Photo courtesy Archer Stephenson

2-55: Photo courtesy M12 Archive, Jeffrey Machtig, John Michael Kohler Arts Center

2-56: Motmit/Wikimedia Commons

2-57: Javierri/Wikimedia Commons

3-5: Photo courtesy Ross Hogarth

4-6: TDC Photography/Shutterstock.com

4-11: Klaus Hiltscher/Wikimedia Commons

4-14: Photo courtesy Alex Kluft Photography

4-15: Tinseltown/Shutterstock.com

4-17: Kate Gabrielle/Flickr Commons

4-18: NBC Television/Wikimedia Commons

4-19, 4-21: Stuart Sevastos/Wikimedia Commons

4-24: Mel Vyvyan/Wikimedia Commons

4-25: Photo courtesy Dani Macchi

4-26: AVRO/Wikimedia Commons

4-30, 5-19, 6-16: Photo courtesy Mark Rubel

4-31: Photo courtesy Justin Stanley

4-36: Christian Bertrand/Shutterstock.com

4-39: Roy Kerwood/Wikimedia Commons

4-43: Back9Network/Flickr Commons

5-8: Photo courtesy Warm Audio
5-9: Magnushk/Wikimedia Commons
5-12: Jim Summaria/Wikimedia Commons
6-10: Wikimedia Commons
6-15: Photo courtesy Peter Schaaf
6-18: Photo by Jeff Brown
6-19: Photo courtesy Sherman Electronics
6-35: Photo courtesy Ross Garfield
6-38: Photo courtesy Trixon
6-42: Photo courtesy Marc Kallweit
7-4, 7-5: Photo courtesy Jason Benham
7-6: Photo courtesy Kevin Jardine
7-31: Photo courtesy Scott Lee
7-33: Kreepin Deth/Wikimedia Commons
7-36: Carl Lender/Wikimedia Commons
8-4: Photo by Jandro Cisneros Photography
8-9: Jack Mitchell/Wikimedia Commons
8-10: Matt Eason/Wikimedia Commons
8-16: Aristide1811/Wikipedia
8-17: Andreas Praefcke/Wikimedia Commons
9-2: Photo courtesy Ocean Way
9-7: Photo courtesy Pete Thomas
9-8: Photo courtesy Steven Jarvis
9-15: Koen Suyk/Wikimedia Commons
9-16: Sara Guastevi/Wikimedia Commons
9-18: Photo courtesy Lori Castro
10-2: Les Chatfield/Wikimedia Commons
10-3: Dennis AB/Flickr Commons
10-4: Brennan Schnell/Wikimedia Commons
10-9: Photo courtesy Julian Colbeck
10-19: Nomo/Michael Hoefner/Wikimedia Commons
11-2: Photo courtesy Stephen Shrimpton
11-4: Joe Mabel/Wikimedia Commons
11-8: Photo courtesy Mark Pauline
11-9, 13-25: Photo courtesy Mark Arinsberg
11-10: Chris Friese/Wikimedia Commons
11-11: Photo courtesy Michael Wilson
12-2: Photo courtesy Bob Ezrin
12-3: Capitol Records/Wikimedia Commons
12-4: Mr. Bonzai/David Goggin ©1993
12-6: Ben Ghazi Enterprises/Wikimedia Commons
12-7: Archive of the Stockhausen Foundation for Music, Kuerten (www.karlheinzstockhausen.org)
12-9: Photo courtesy Preetam Slot/Flickr Commons
12-15: Photo courtesy Annabel Tehran
12-21: Photo courtesy Jack Joseph Puig
12-24: Dana Nalbandian/Shutterstock.com
12-27: Photo courtesy Gustavo Pereyra
13-5: Photo by Jason Quigley, courtesy Larry Crane
13-6: Images courtesy Globe Institute of Recording and Production
13-12: Bravedeer/Wikimedia Commons
13-21: Photo courtesy Pete Weiss
13-22: Photo by Victor Levy-Lasne
13-23: Anna Wittenberg/Wikimedia Commons